Common Interior Alaska Cryptogams

Fungi, Lichenicolous Fungi, Lichenized Fungi, Slime Molds, Mosses, and Liverworts

Common Interior Alaska Cryptogams

Fungi, Lichenicolous Fungi, Lichenized Fungi, Slime Molds, Mosses, and Liverworts

Gary A. Laursen

and

Rodney D. Seppelt

Contributing Authors:

Mikhail P. Zhurbenko, Steven L. Stephenson, Harold H. Burdsall, Linda H. Geiser, and Roseann V-E Densmore

University of Alaska Press

Fairbanks, Alaska

University of Alaska Press
P.O. Box 756240
Fairbanks, AK 99775-6240

ISBN 978-1-60223-058-3

Library of Congress Cataloging-in-Publication Data

Laursen, Gary A.
Common Interior Alaska cryptogams : fungi, lichenicolous fungi, lichenized fungi, slime molds, mosses and liverworts / Gary A. Laursen and Rodney D. Seppelt.
p. cm.
ISBN 978-1-60223-058-3 (pbk. : alk. paper)
1. Cryptogams—Alaska—Interior Alaska. I. Seppelt, R. D. II. Title.
QK511.A4L38 2009
586.09798—dc22
2009005577

Photos by Gary Laursen, Rod Seppelt, Mikhail Zhurbenko, Steve Stephenson, Jim Riley, and Michael Lueth; line and schematic drawings were made by Maggie Hallam and Rod Seppelt.

Cover design by Dixon Jones, Rasmuson Library Graphics

Text design by Paula Elmes, ImageCraft Publications & Design

This publication was printed on acid-free paper that meets the minimum requirements for ANSI / NISO Z39.48-1992 (R2002) (Permanence of Paper for Printed Library Materials).

Printed in the United States of America

Preface

The natural world is exceedingly diverse. While most people are aware of the landscape, few ever fully comprehend or appreciate the vast diversity of color, shape, size, and morphology of the component organisms, and diversity of their interrelationships with other organisms or the environment in which they live.

Interior Alaska is a vast and diverse area of wilderness set between two large mountain ranges, the Brooks Mountains to the north and the Alaska Range to the south. Nestled nearer to the latter is Denali National Park and Preserve. It includes North America's highest mountain, permanently snow-covered peaks, glaciers, rivers and streams, lakes, a diverse geology and vegetation that ranges from alpine tundra to boreal forest and wetlands, and a rich flora and fauna. Many thousands of visitors come to Interior Alaska each year to experience wilderness, to appreciate the diverse landscape and natural beauty, and to glimpse the wildlife. Few, however, come to appreciate the beauty and diversity of the cryptogamic flora or ever gain an understanding of the importance and role of these groups of organisms in the ecosystem.

With this field guide, we aim to redress this deficiency by providing an introduction to these lesser-known or appreciated organisms in a way that will be informative to visitors to Interior Alaska.

Contents

List of Figures

Fungal Group: The Hymenomycetes (Having an Organized Hymenium)

Fungal Group: The Gasteromycetes (Having No Organized Hymenium)

Fungal Group: Conifer Rust, Jelly, Cup, and Earth Tongue Fungi

Plasmodial Slime Molds (Mycetozoans)

Bryophytes: The Mosses, Liverworts, and Hornworts

Figure 1. Spectacular mountain ranges

Introduction

For years visitors have come to the land of midnight sunsets, spectacular mountain ranges (Fig. 1), and vast pristine expansive vistas—some marred by mining disturbances (Fig. 2) and others under restoration (Fig. 3). It is the land of sweeping boreal forests pocked with small lakes being choked by floating moss mats (Fig. 4), where spruce reigns supreme. Views of mountain landscapes, forests "cloned" by a mosaic of colorful patterns (Fig. 5), tundra meadows with rock scree polygons (Fig. 6), and riverbeds reached only by helicopter transport (Fig. 7) are often nothing less than awe-inspiring, but there are also many other and unusual life-forms present. To the eyes of off-road hikers and other seekers of a wilderness experience, mushroom fungi, lichenized fungi, lichenicolous fungi, slime molds, mosses, and liverworts (Fig. 8a–f) abound during their fruiting period of about six weeks that generally extends from mid-July to late August. Fall comes early to these northern subarctic regions, and with an abundance of rain, cool night temperatures, and plant-derived nourishment, cryptogamic organisms becomes abundant. Much of the effort that has gone into the preparation of this treatment is predicated on work performed in Denali National Park & Preserve.

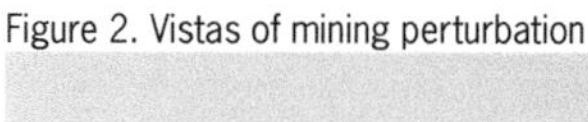
Figure 2. Vistas of mining perturbation

Figure 3. Vistas under restoration

Figure 4. Sweeping boreal forests

Figure 5. Forests cloned by a mosaic of colorful patterns

Figure 6. Tundra meadows with rock scree polygons

Figure 7. Riparian riverbeds accessible only by helicopter transport

Figure 8a. Mushroom fungi

Figure 8b. Lichenized fungi

Figure 8c. Lichenicolous fungi (*Nectriale canodes* on *Peltigera*)

Figure 8d. Slime mold plasmodium

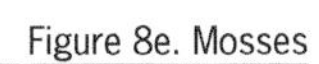

Figure 8e. Mosses

Figure 8f. Liverworts

Figure 9. Denali National Park entrance

Interior Landscapes

The Setting

Significant Interior Alaska landscapes are included in one of our nation's largest parks. On February 26, 1917, President Woodrow Wilson signed legislation that created Mount McKinley National Park, renamed Denali National Park and Preserve in 1980 (Fig. 9). In 1922, the park was enlarged to the east and in 1932, to the west. Denali State Park, an area of 324,240 acres, was created in 1970. Lands within the National Park were designated as "wilderness." The addition of "preserve" lands to the existing "park" lands, in 1980, tripled the size of what then became Denali National Park & Preserve, hereafter referred to as DNP&P, to its present 6,028,091 acres (Figs. 11 and 12). Mount McKinley (Fig. 10), called **Traleika** (Athabascan for "high mountain") and later **Denali** (Athabascan for "great one"), was in later years referred to as Densmore's Mountain after prospector Frank Densmore, who frequented the

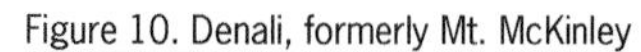

Figure 10. Denali, formerly Mt. McKinley

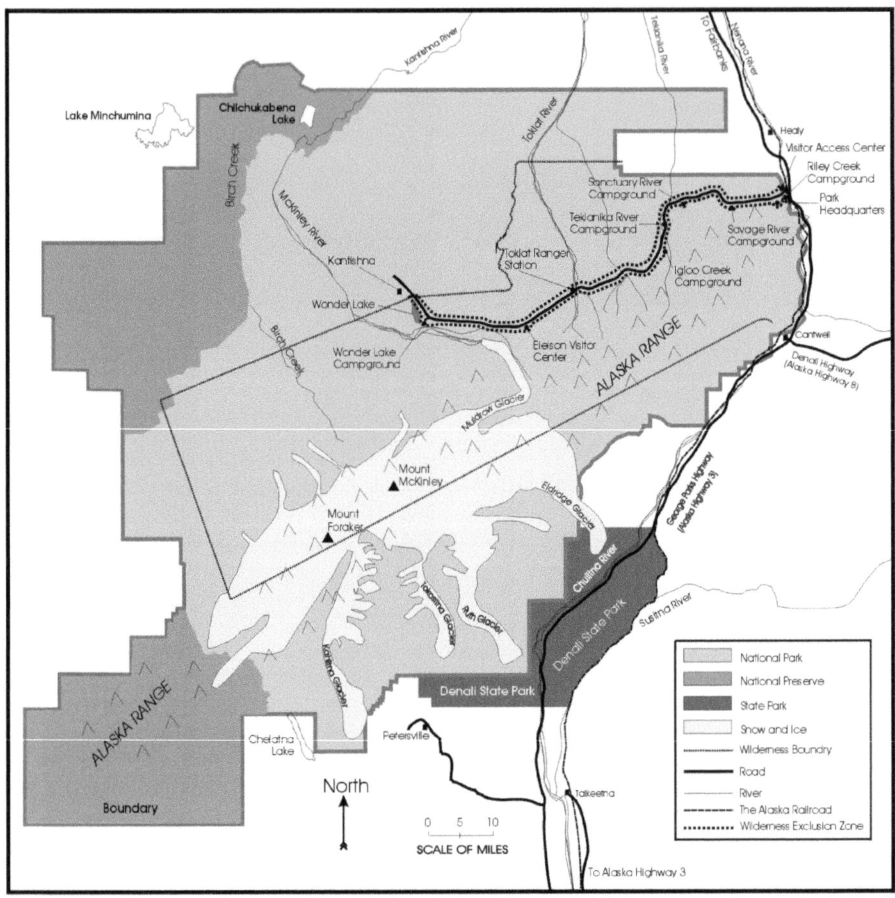

Figure 11. Denali basemap. Courtesy of Denali National Park and Preserve, U.S. Department of the Interior, National Park Service.

area. The name "**McKinley**" is derived from the name of Ohio governor William McKinley (1843–1901), who became our nation's twenty-fifth president. It is a subarctic mountain where the sun sets on June 21 at 11:30 p.m. and rises again at 12:45 a.m. On December 21, the sun rises at 9:30 a.m., but sets at 3:00 p.m. In 1913, Walter Harper became the first to climb Mount McKinley. The mountain is within the taiga ("land of little sticks") biome and is known for its upper boreal forest components of white and black spruce and alpine and subarctic tundra ("land of no trees"), with precipitous landscapes, solifluction (Fig. 13), and meandering rivers and streams. The setting is a geomorphologist's opened textbook for all to witness.

DNP&P is dominated by lofty, snow-covered mountain peaks constituting, in part, the Alaska Range (Fig. 14), a section of the Alaska Aleutian Physiographic Province of the Pacific Mountain System. Portions of the northern foothills of the Alaska Range are in the central and eastern sections of DNP&P, while the western section occupies the Tanana-Kuskokwim Lowlands of the Western Alaska Province. Elevations range

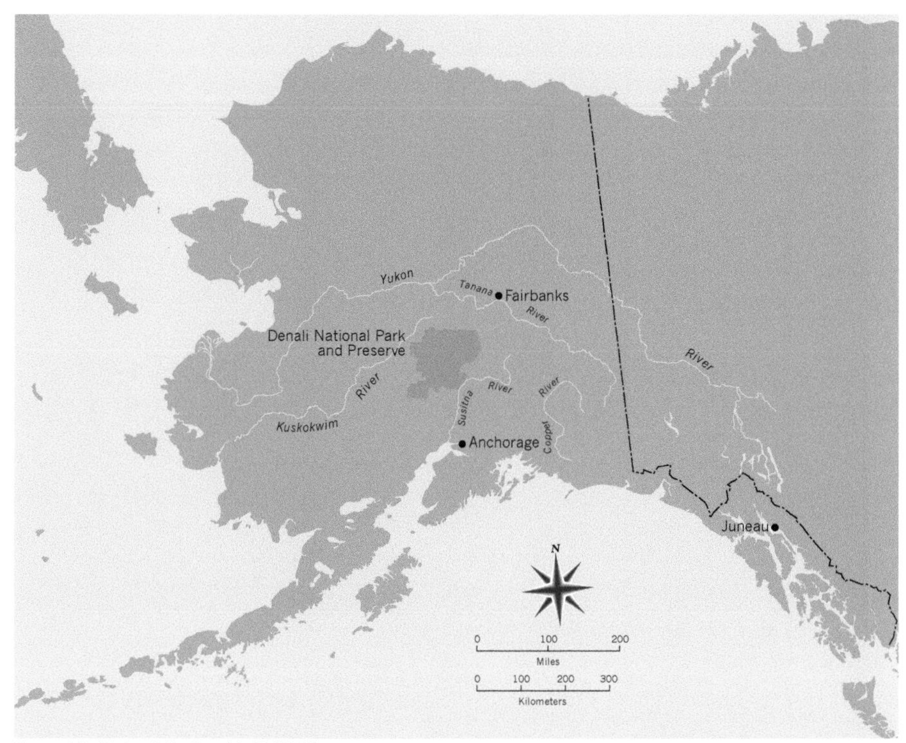

Figure 12. State of Alaska with DNP&P insert

Figure 13. Precipitous landscapes support solifluction

Figure 14. Lofty, snow-covered mountain peaks of the Alaska Range

Figure 15. Spruce forests dissected by braided, meandering to incised streams

from 122 m (400 ft) to 6,194 m (20,320 ft) at the top of Mount McKinley, the highest peak in North America and the western terminus of the Alaska Range. Six major glaciers 40 to 72 km (25 to 45 miles) long run away from the mountain, leaving glaciated U-shaped valleys as they melt and from which silt-laden waters flow year-round. The northwestern section, a relatively flat area of lowlands covered by white spruce forests, contrasts strikingly with the mountainous areas. About half of the Alaska interior consists of mountains and ridges above 1,219 m (4,000 ft) elevation. Above 2,134 m (7,000 ft), snow and glaciers remain continuously in these subarctic mountains where elevations of 1,829 to 2,134 m (6,000 ft to 7,000 ft) are common. Average heights increase as one approaches Mount McKinley, with many peaks exceeding 3,048 m (10,000 ft). North of the Alaska Range, the landscape encompasses areas of rolling tundra. These rise into hills and a system of less-than-rugged mountains called the Outer Range. The landscape then sweeps into low-lying boggy muskeg and omnipresent boreal spruce forests dissected by braided, meandering to incised streams, creeks, and brooks (Fig. 15). With few roads, but hosting myriad trails and footpaths, the land is bisected by a single 146 km (91 miles) well-maintained gravel road running east to west to the south of the Outer Range and running out into the gold-rich Kantishna Hills. Here is where a great deal of the work on cryptogams in the Alaska interior has been focused.

Climate

The Alaska interior lies within two major climatic zones. A continental climate exists to the north of the Alaska Range, whereas a transitional climate (ranging from coastal maritime to continental) is exhibited to the south. The north is characterized by 381 mm (15 in) of precipitation annually, with 1,905 mm (75 in) of snow and greater fluctuation in temperature (hotter in summer and much colder in winter) than the area to the south. About half of the summer days are cloudy, and Mount McKinley is not visible, or only partially visible. Clearer skies persist with cooler temperatures in fall, winter, and spring. Summer temperatures generally range from 1.7° to 18.9°C (35° to 66°F), but may occasionally reach 32°C (90°F). Winter temperatures range from –21.7° to 2.8°C (–7° to 27°F) and may drop to –47°C (–52°F). Winter temperatures at higher elevations are generally more severe. The Alaska Range is a barrier to north-south air movements and precipitation from the maritime influences to the south, thus creating a transitional climate. Areas on the south side of the range receive precipitation amounts at least twice those measured on the north side. There are more cloudy days on the south side throughout the year, largely as a result of storms that develop in the Gulf of Alaska and are pushed up against the range. Temperatures on the south side of the range exhibit less variation and tend to be warmer in winter and cooler in summer. Mean temperature ranges near park headquarters at the east entrance to the park are –20.6° to11.0°C (–5.2° to 12.0°F) in January and 6.5° to 21°C (43.8° to 69.8°F) in July.

The average annual precipitation at DNP&P headquarters is slightly more than 36 cm (14 in); snowfall there is 192 cm (75.7 in). At higher elevations in the Alaska Range, precipitation exceeds 203 cm (80 in) and snowfall exceeds 1,015 cm (400 in). Normal snowpack throughout the region averages between 51 and 102 cm (20 and 40 in).

Physiography

Interior Alaska is dominated by Mount McKinley and a series of two and sometimes three parallel, rugged, and glaciated mountain ridges known as the Alaska Range. High mountains (Mount McKinley at 6,194 m (20,320 ft), Mount Foraker at 5,304 m (17,400 ft), and Mount Hunter at 4,442 m (14,573 ft) within the range are perpetually snow-clad generally above 2,134 m (7,000 ft), where numerous glaciers originate. Northern foothills of the Alaska Range consist of a series of east-westerly oriented ridges, starting with the Kantishna Hills and running eastward. Summit altitudes generally range from 610 to 1,372 m (2,000 to 4,500 ft). Ridges are separated by broad glacial valleys that range from 3 to 16 km (2 to 10 miles) in width. Southern foothills are generally quite steep and are cut through by large south-flowing glaciers. To the south of the Alaska Range, and west of the Talkeetna Mountains, broad Susitna River lowlands reach out in a north/south direction, where elevations are less than 152 m (500 ft).

Fungi: Mushrooms and Other Cryptogams

In the vast subarctic landscape that is interior Alaska, more than 250 species of vascular plants share their habitat with an extraordinary abundance of cryptogams.

Most people who visit have a general idea of what fungi are, that they occur in many different habitats, and that great care should be taken in any attempt to collect and prepare them for the table.

> **REMEMBER!** Collections of any plant or animal material within Denali National Park & Preserve are prohibited by law and can only be made in the company of research applications and issued permits!

The most familiar examples of cryptogams are the fungi, which include the Basidiomycetes (club fungi) that include agaricoid (gilled mushrooms), boletinoid, polyporoid, toothed, coraloid, cantharelloid, thelephoroid, gasteroid (puffballs and bird's nests), rusts, jellies, and Ascomycetes (sac fungi), the cup fungi, and earth tongues. There are also the less familiar cryptogam groups that include the lichenicolous fungi, lichenized fungi or lichens, and more distantly related groups such as slime molds (*Perichaena minor*) (Fig. 16a–q). Also included in the cryptogams are the mosses and liverworts, photosynthesizing plants that are very different from the fungi. Few people, however, know very much about fungi or other cryptogams. Those who study fungi are referred to as mycologists, whether amateur or professional, while those studying the mosses and liverworts are bryologists.

Fungi (singular: fungus) are unique organisms, so much so that they have been assigned by taxonomists (i.e., scientists who name and classify organisms) into their own kingdom—the Fungi (Myceteae). Their sudden appearance (which requires rains or moisture), vivid colors, lack of chlorophyll (the photosynthetic pigment of higher plants, requiring the fungi to find a suitable carbon source as nutrition for survival), unusual odors, delicate flavors (of the edible kinds), varied life-forms (shape or habit), and distinctive textures (fleshy, dry to moist, tacky to sticky, gelatinous, slimy to outrageously gooey) all add to their mystery and attraction. Many people are fascinated by the mystique surrounding the edibility, toxic poisoning, and hallucinogenic properties of fungi.

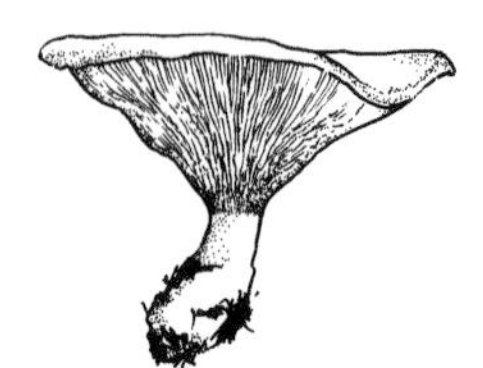
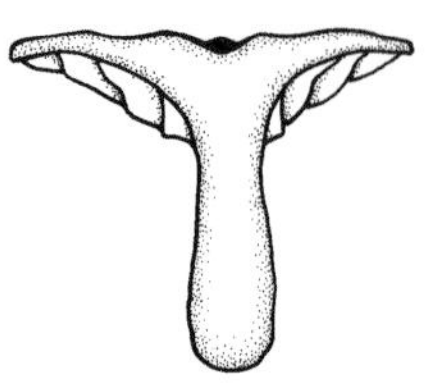

Figure 16a. Agaricoid

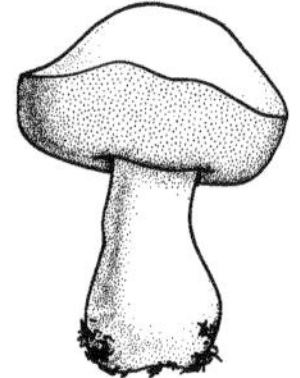

Figure 16b. Boletinoid

Figure 16c. Polyporoid

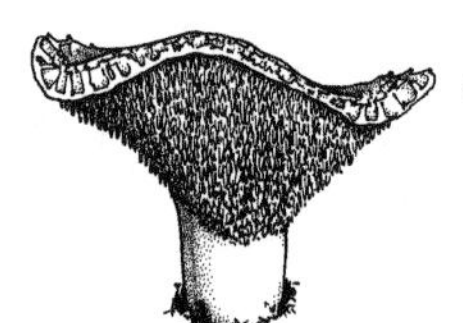

Figure 16d. Toothed fungi

Figure 16e. Coraloid fungi

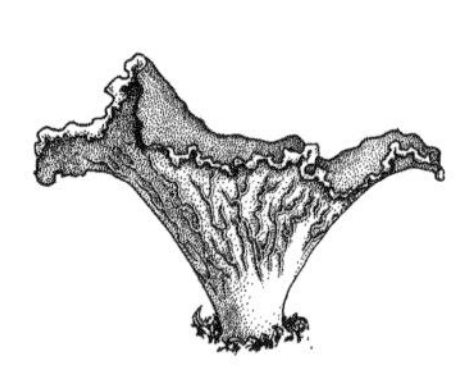

Figure 16f. Cantharelloid

Figure 16g. Thelephoroid

Figure 16h. Gasteroid puffball

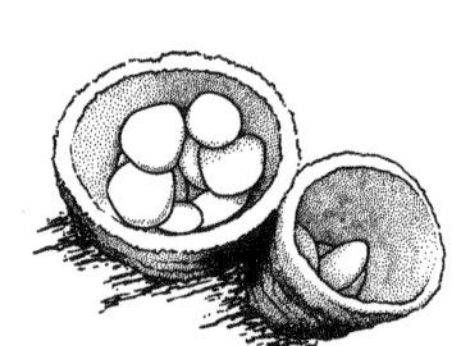

Figure 16i. Gasteroid bird's nest fungi

Figure 16j. Earthstar

Figure 16k. Conifer rust fungi

Figure 16l. Jelly fungi

Figure 16m. Cup fungi

Figure 16n. Earth tongue fungi

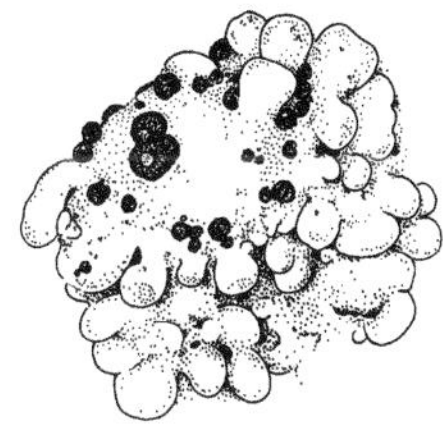

Figure 16o. Lichenicolous fungi

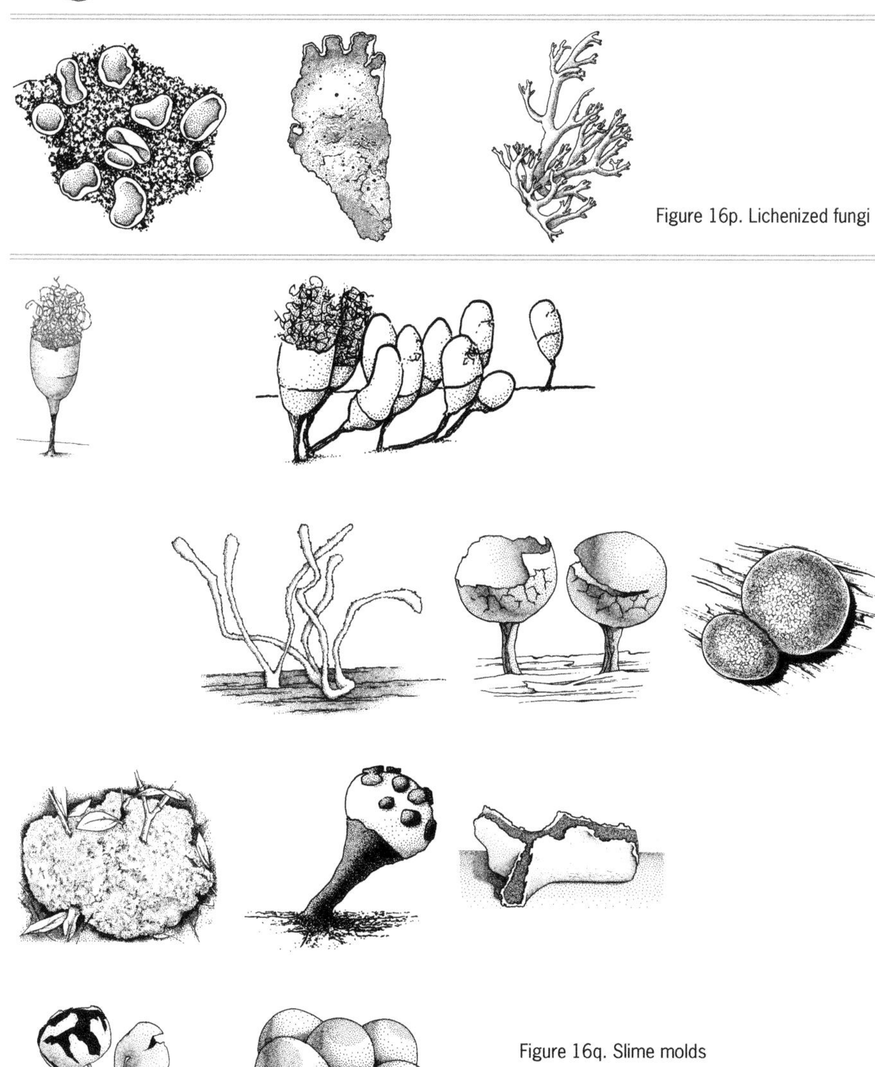

Figure 16p. Lichenized fungi

Figure 16q. Slime molds

Fungi are absolutely necessary for the dynamic balance that characterizes the natural world. About fifty percent of Interior Alaska fungi are decomposers (saprophytes). Decomposition can be slow in the Alaska interior and in more northern latitudes, where growing and decomposing seasons are significantly shortened by a cooler climate, but it does happen and the evidence is everywhere.

As well as more easily seen fungi, this field guide includes those other groups of cryptogamic organisms, the **lichenized fungi** (an intimate symbiotic association of fungus with either green algae or cyanobacteria forming the lichen), and the **lichenicolous fungi** (fungi that are parasitic or saprophytic on lichens). We have also included the **slime molds**, once included in the fungi Myceteae but now classified in an entirely separate group, the Protista. We also include the cryptogamic mosses and liverworts, complex but nevertheless fascinating groups that belong to early diverging lineages in the kingdom Plantae.

Fungal Ecology

Found almost everywhere, fungi occupy myriad habitats, from terrestrial to aquatic (fresh, brackish, marine, glacial water, and permafrost). They occur on a wide variety of substrates including, but not limited to, bone; hair; old leaves; logs; assorted woody fiber, chips, and paper; cloth and leather fabrics; mosses; soil; plant roots; the lenses of cameras; and even in jet fuel! However, fungi must have moisture to live in any environment. Some fungi need sunlight, not to make food as green plants containing chlorophyll do, but to stimulate their fruiting response. They produce carbon dioxide (CO_2) gas by respiration, much as humans do.

Most fungi live on dead organic matter (often referred to as **detritus** or **humus**) and, therefore, are termed **saprobes**, **decomposers,** or **saprophytes**. As such, they are involved in a process (decomposition) that ensures the release of nutrients back into the environment. Other fungi thrive on living plants, animals, and also other fungi, and are called **parasites** (mycoparasites, lichenicolous, rust, smut, jelly, agaric, and polypore fungi). Still others live in intimate, mutually beneficial associations with microscopic and most often single-celled green and blue-green algae (the cyanobacteria) and in this **symbiosis** (living together) they are called **lichenized fungi**, or simply **lichens**. Others play a different, but equally significant, role in nature. Without them the Alaska interior would be devoid of all of its trees and shrubs, most of its beautiful flowering plants, many of its grasslike forms, and even some of its mosses. Trees, shrubs, and most flowering plants—especially orchids, the seeds of which die if they cannot immediately establish contact with an appropriate fungus at germination—entering into a **symbiotic** relationship that involves their own roots (Latin: "**rhiza**") and the underground microscopic threads (**hyphae**) of a fungus (Greek: "**myco**"). These symbioses are called **mycorrhizae** (fungus-roots) and demonstrate many kinds of mycorrhizal relationships (Fig. 17a–e).

Many insects also depend upon fungi for their livelihood and the completion of their own life cycles. Insects often get to fungi well before we do and deposit their

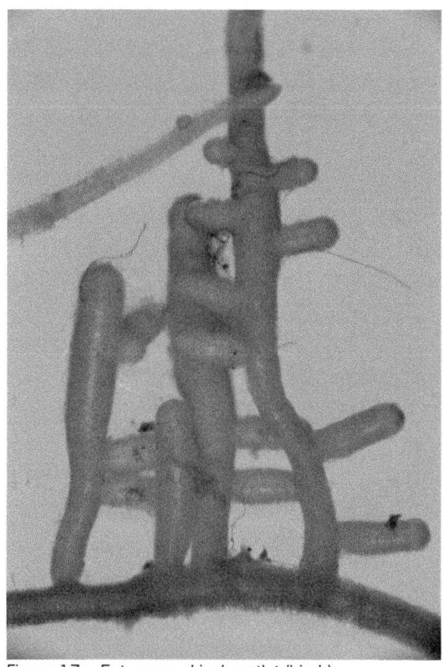
Figure 17a. Ectomycorrhizal rootlet (birch)

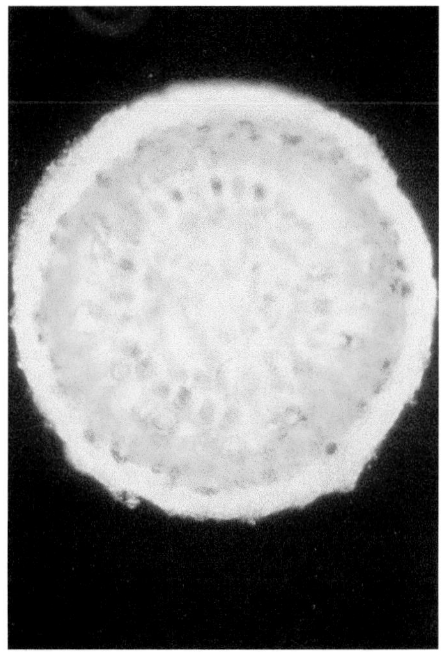
Figure 17b. Ectomycorrhizal rootlet cross section (willow)

Figure 17c. Ectomycorrhizal rootlet mantle
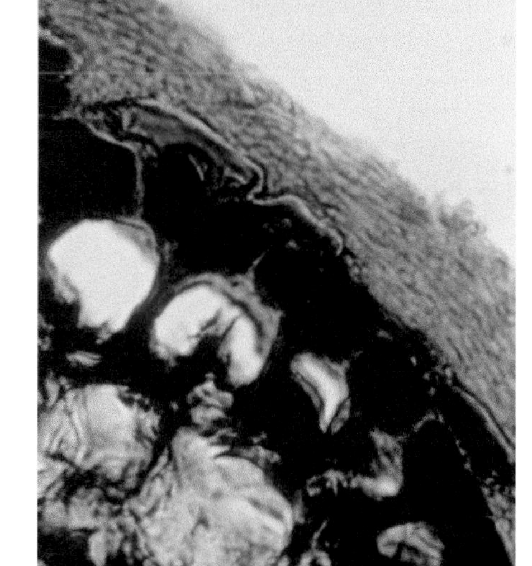

Figure 17d. Mantel mycorrhizae (alder)
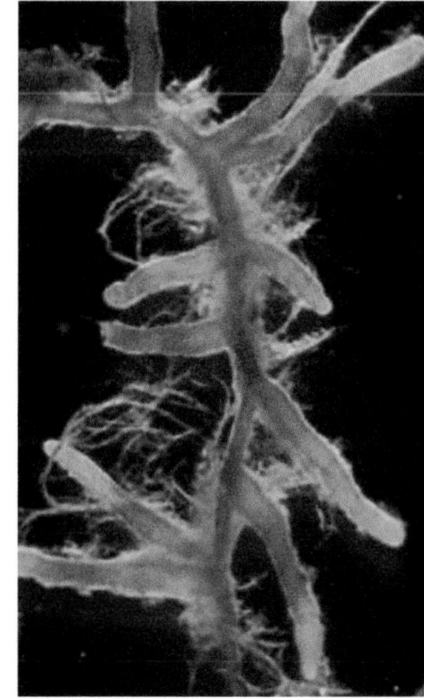

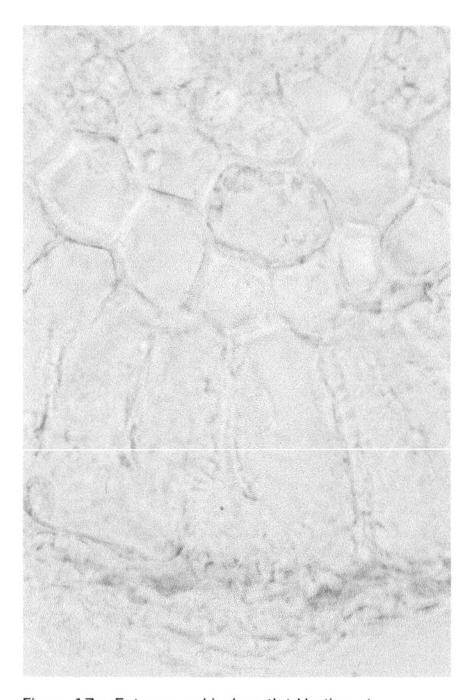
Figure 17e. Ectomycorrhizal rootlet Hartig net

eggs directly onto the spore-bearing surfaces and tissues (**hymenium**) of a fruiting body. If you pick a mushroom up and cut it in half, you may see little tunnels, often stained, running through the tissues. You may even observe squirming maggots (fly and gnat larvae) feeding on the fungal flesh. If you pick up a gilled mushroom and tap it lightly over the palm of your hand, often this will reveal hoards of tiny gray insects called springtails (Collembola) that graze for spores on the gill-like plates of the mushroom. If you pick a corky, leathery to woody bracket fungus from an old log and break or cut it open, you are likely to find a number of tiny beetles. And even mammals get in on the fungal smorgasbord. Caribou and moose may selectively pick and eat fungi right out in front of you. Voles delight in their savory flavors and nutrition. Squirrels air dry mushrooms in the crotches of branches and then place them into old nests high in the crown of spruce trees, often in a mass of branches (a witches' broom) still living or perhaps dead, but once infected with a rust fungus (*Chrysomyxa arctostaphylli*). Witches' brooms are caused by hormones produced by the rust fungus and others, including *Taprhina* spp. You might see these masses of branching limbs as large orange-colored balls high in the crowns of white spruce (*Picea glauca*) and black spruce (*Picea mariana*) trees. The cycles in the forest (Fig. 17f) involve three mammals, mycophagy (eating of fungi), and numerous species of fungi that demonstrate all four principal roles, which are again played by them through time.

Mushroom Architecture

Part of what makes identifying fungi or other "botanical" organisms difficult for the casual observer is the overwhelming number of terms used by so many authors in field guides like this to describe both their macro- and micromorphological features. In our attempt to assist you in the pursuit of understanding these wondrous organisms more fully, we thought it best to assist the visual learner in most of us by providing schematic diagrams to depict many of the terms we use in describing the 209 organisms pictured here. You may find it very useful to glance through Figures

Interactions of the Mammal Mycophagy and Mycorrihizae Cycle

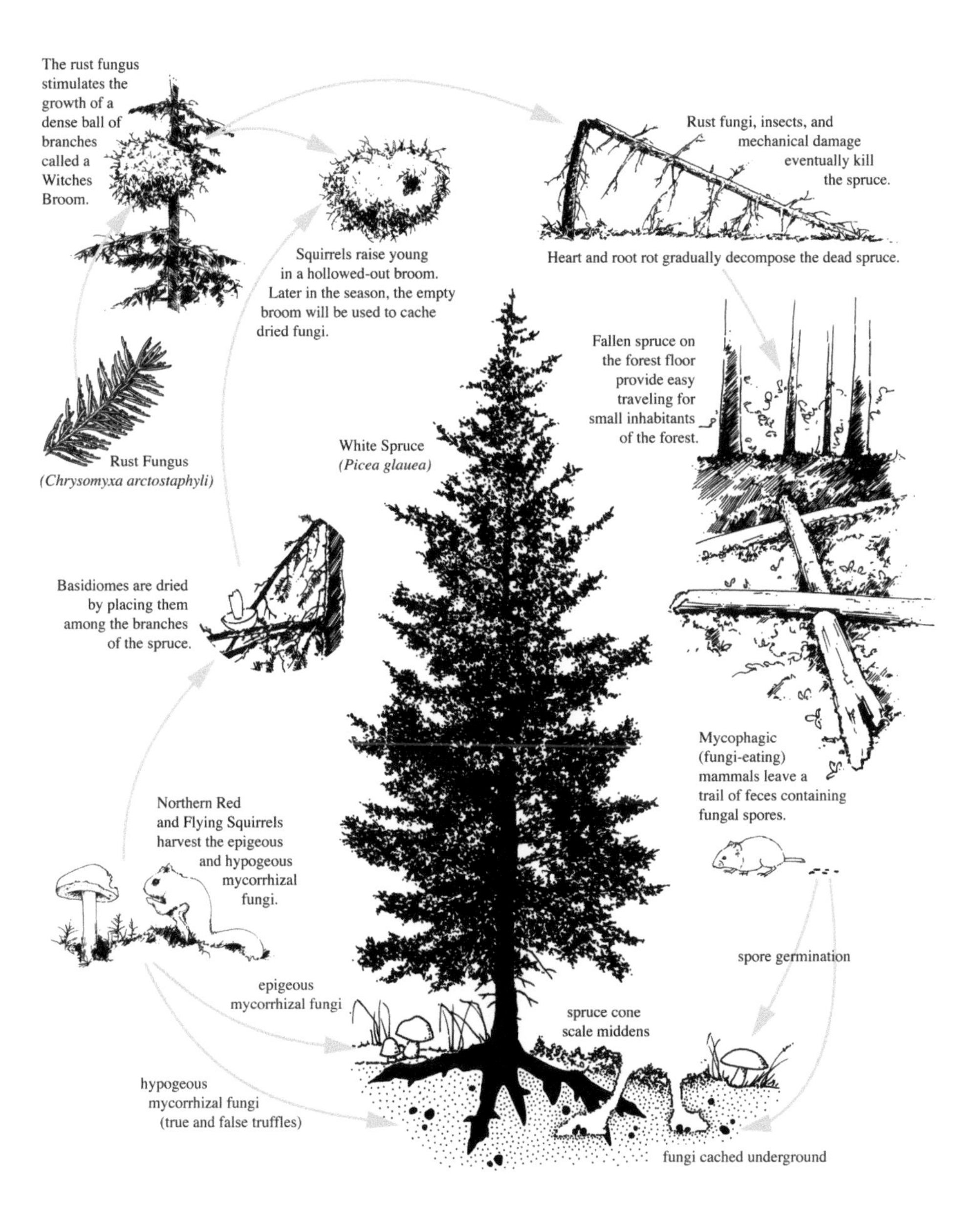

Figure 17f. Cycles in the forest

18–38 to gain perspective and an appreciation for the vast variation in macroscopic architectural morphology of fungal "parts."

A flowering plant consists of leaves, stem(s), and roots (collectively representing its vegetative parts) and various floral parts used for sexual reproduction. A similar situation applies to fungi. The vegetative or "assimilative" (water- and nutrient-gathering, or literally "eating") part is called the **mycelium** and consists of a mass of tiny cottonlike microscopic threads called **hyphae** that permeate or cover the substrate upon or within which the fungus grows (Fig. 18). These mycelial masses are composed of many interconnected hyphae that are not easily seen. We call the reproductive structure—the mushroom itself—a fructification (its use now discouraged because fruitbodies are not homologous structures), basidiocarp, basidiome, ascocarp, ascome, or simply the **fruiting body** that often demonstrates a cap (pileus), hymenium (gills, teeth, pores, etc.), a stipe (not a stem), and an often differentiating base (Fig. 19) if present.

Figure 18. Fungal mycelium consisting of a mass of tiny cottony, microscopic threadlike hyphae

For instance, Figure 20a–f depicts not only basic mushroom architecture but also a maturation sequence. Figure 21a–d takes us a bit further to learn about basidiomycete architecture and spore production, and Figure 22a–d demonstrates the events ascribed to ascomycete architecture and spore production. Figure 23a–f shows us only a fraction of the fungal spore morphologies seen in basidiospores, and Figure 24a–e similarly shares the fungal spore morphology seen in a few ascospores. Figure 25a–e provides fungal spore morphology in the form of spore ornamentation, features often used to discern and describe a fungus at the level of its genus and specific epithet (species). There are vast differences in fungal cap (pileus) morphology (Fig. 26a–w), basidiome pileus margin morphology (Fig. 27a–h), basidiome pileus surface morphology (Fig. 28a–l), basidiome pileus surface ornamentation morphology (Fig. 29a–v), and basidiome pileus surface hairiness morphology (Fig. 30a–i). In contrast to pileus (cap) structures, there are similar contrasting differences in stipe (stalk)

Figure 19. "Typical" fruiting body

morphologies as seen in stipe placement (Fig. 31a–d), shape and base (Fig. 32a–n), the volva (Fig. 33a–g), stipe surface ornamentation (Fig. 34a–l), its interior (Fig. 35a–f), and partial veil remnants (Fig. 36a–j). Even the morphology of the hymenium (hymenophore) show significant and taxonomically important features in spore-producing surfaces (Fig. 37a–o), lamellar (gill) attachment (Fig. 38a–k), lamellar (gill) breadth and shape (Fig. 39a–j), and in lamellar (gill) edges (Fig. 40a–g), in addition to poroid (Figs. 16c and 37a–d), hydnoid (Fig. 16d), and smooth (Fig. 16g) hymenial types.

Figure 41. Mycelia, most often seen as white "fans"

Thus, picking a mushroom will not necessarily damage or kill the "fungus" involved because the actively growing part (vegetative mycelium) of the fungus remains behind in the substrate. In no way, however, is the mushroom a true "fruit." The fruiting body serves a similar function by acting as the site where **spores** (not seeds) are produced, much like an apple is the fleshy structure in which real seeds are produced, although macroscopic seeds and microscopic spores are vastly different. The macroscopic fruiting body is the sexual reproductive part of the fungus and is nourished and supported by its microscopic mycelium. Peel bark from a dead tree and one can often see mycelia en masse, and mostly as white "fans" (Fig. 41), but in Alaska the mycelium may also be tan, yellow, orange, red, or brown. Dig beneath that next mushroom you find and you, too, may get a glimpse of the mycelium composed of rootlike strands called **rhizomorphs** that fan out onto and over suitable

Figure 20a–f. Development and Parts of a Gilled Mushroom

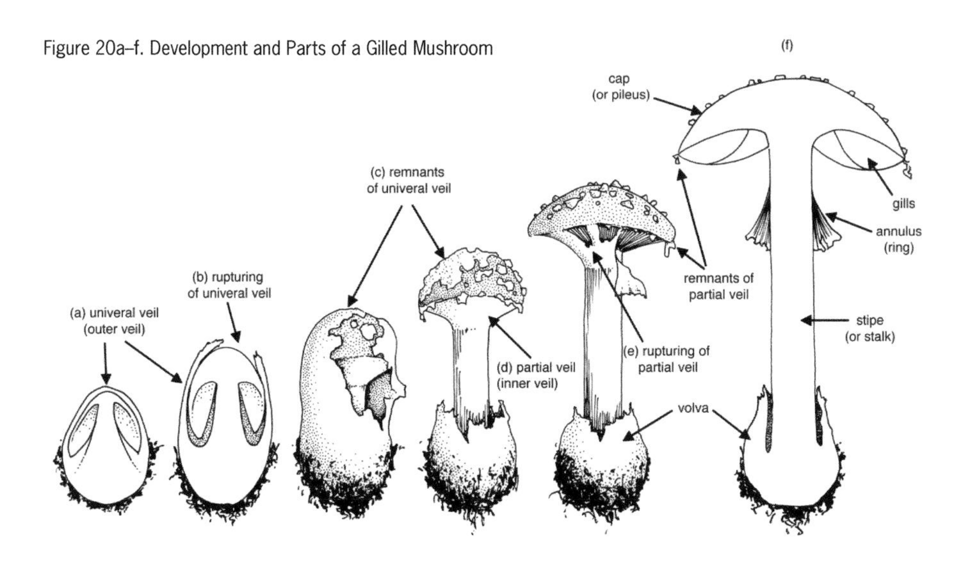

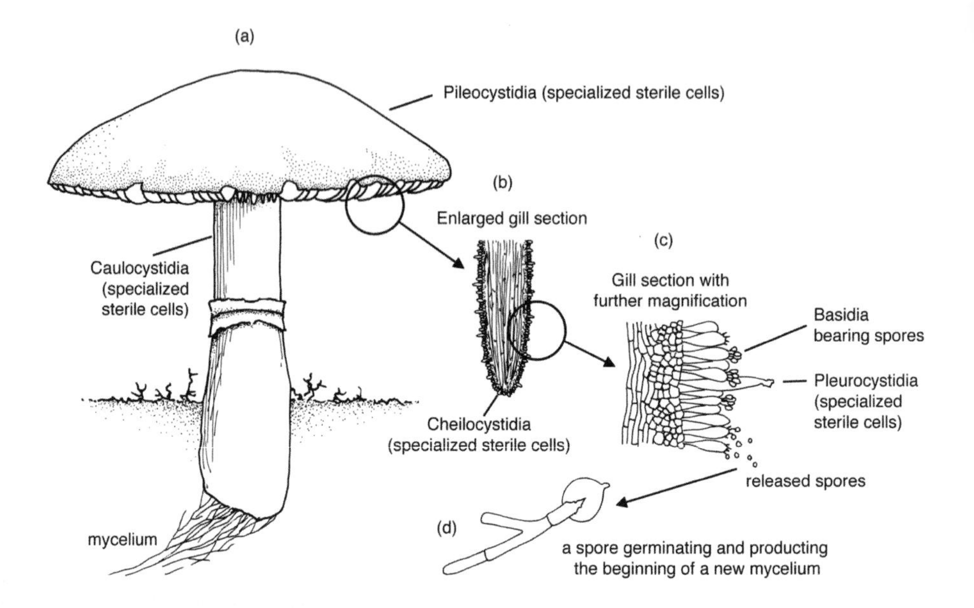

Figure 21a–d. Basidiomycete architecture and spore production

Figure 22a-d. Ascomycete architecture and spore production

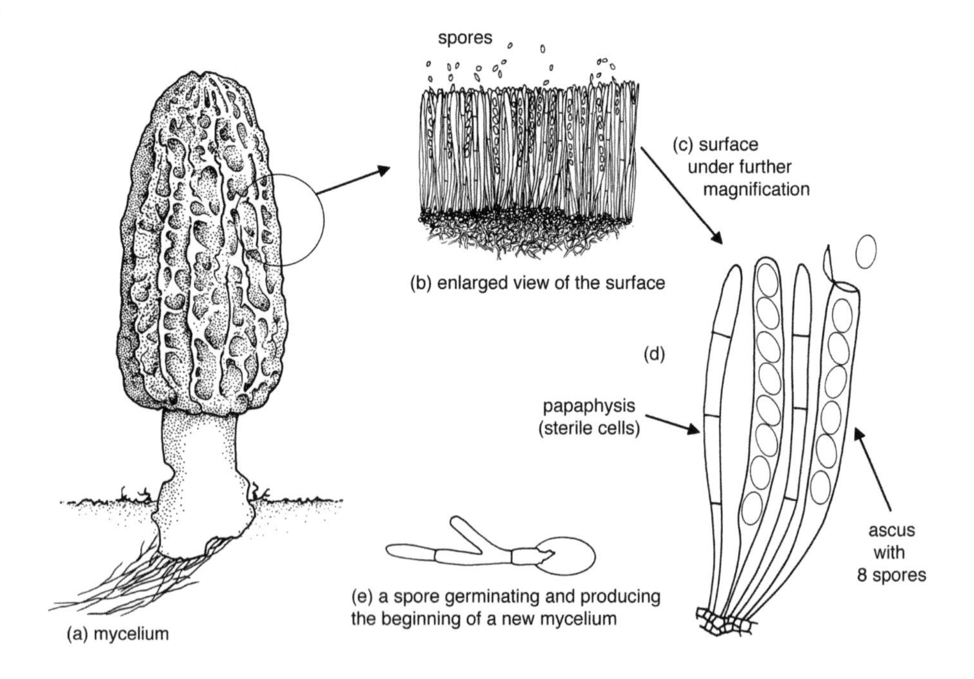

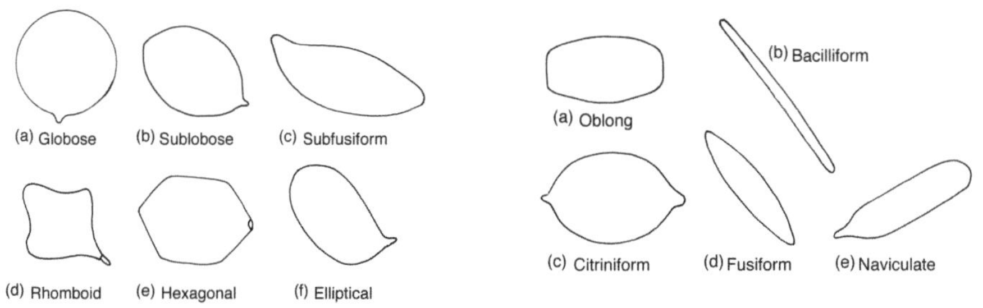

Figure 23a–f. Fungal spore morphology: basidiospores

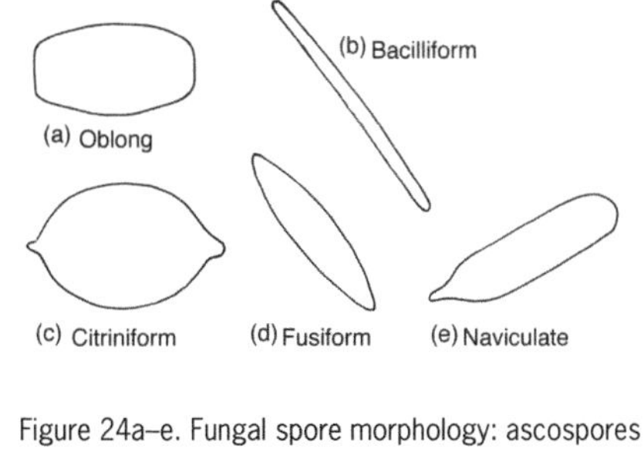

Figure 24a–e. Fungal spore morphology: ascospores

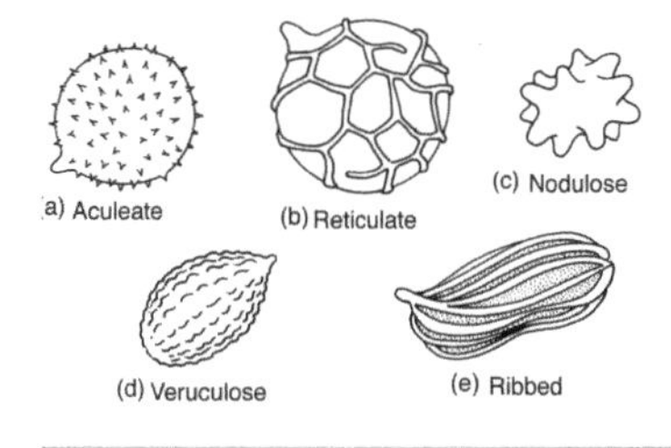

Figure 25a–e. Fungal spore morphology: ornamentation

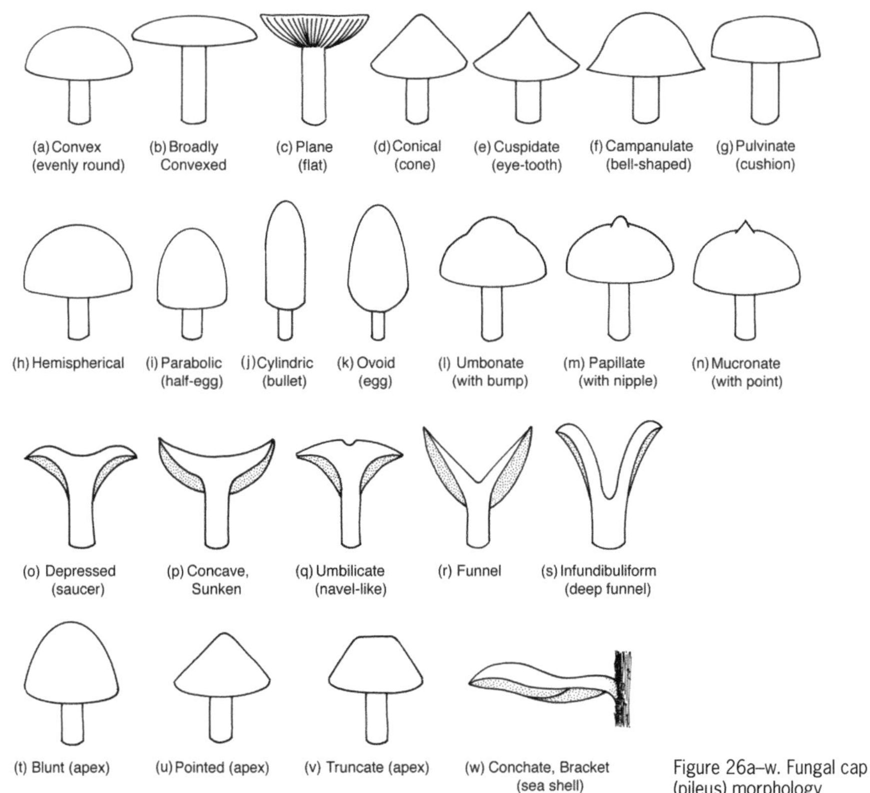

Figure 26a–w. Fungal cap (pileus) morphology

Figure 27a–h. Basidiome pileus margin morphology

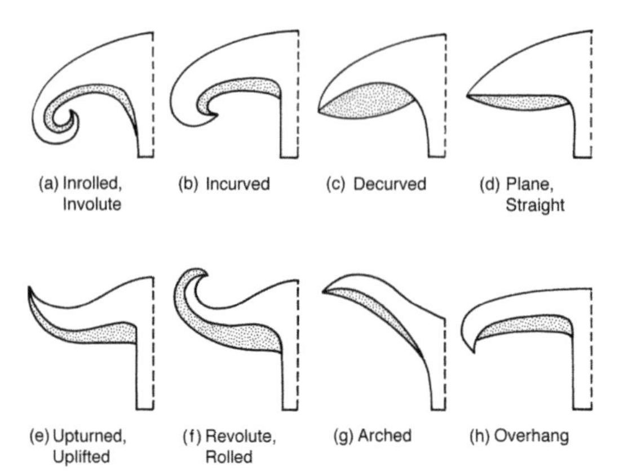

Figure 28a–l. Basidiome pileus surface morphology

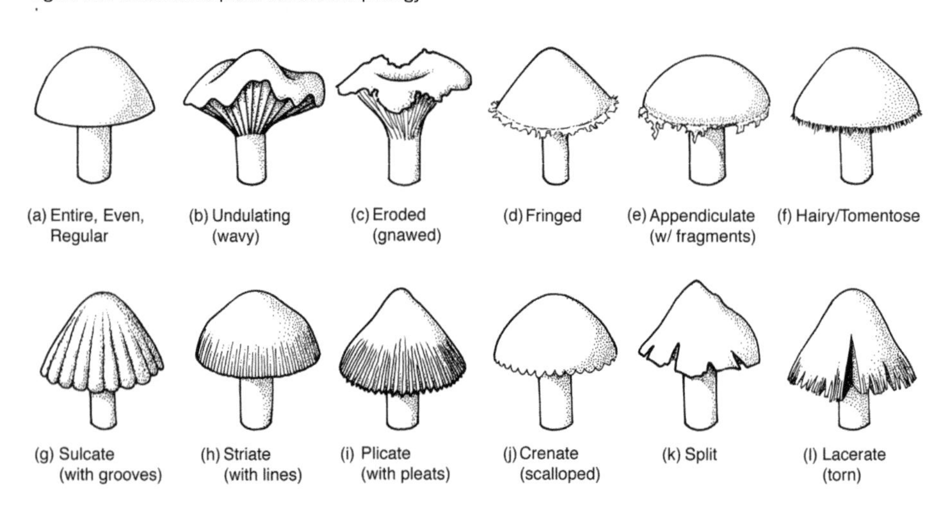

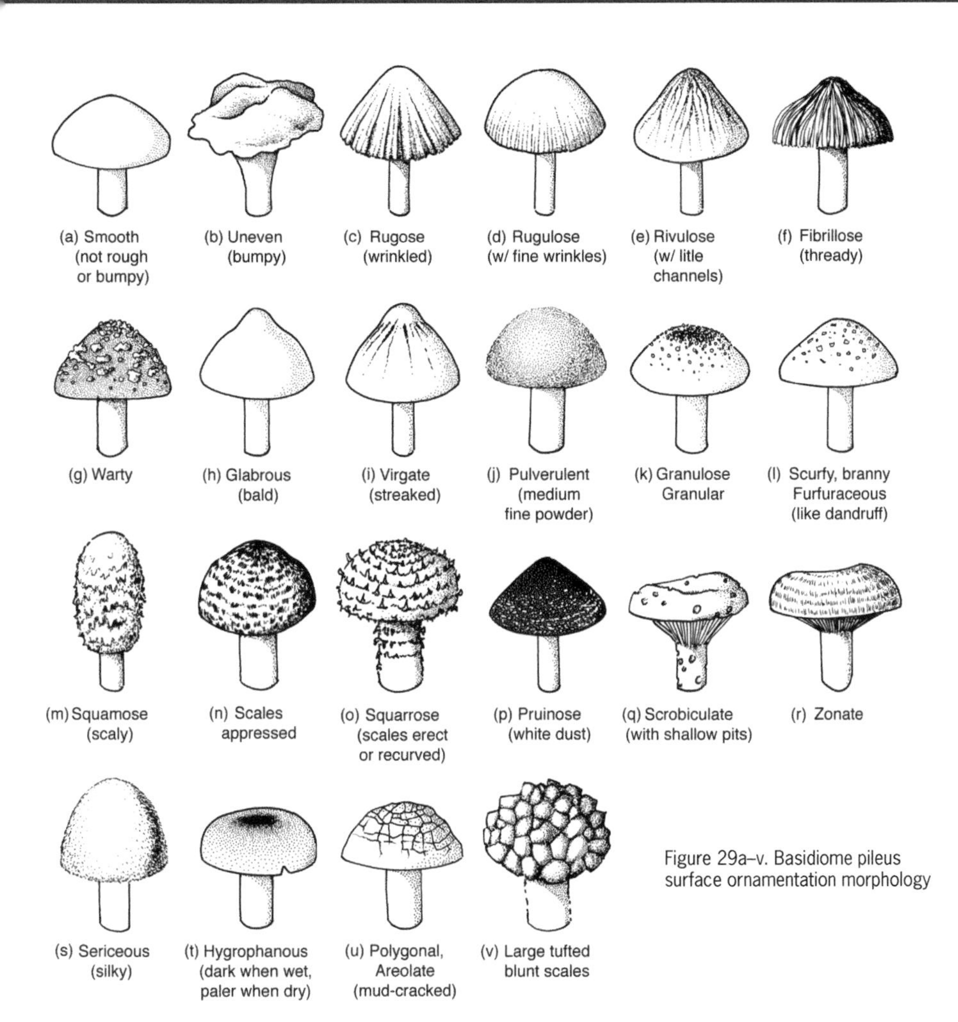

Figure 29a–v. Basidiome pileus surface ornamentation morphology

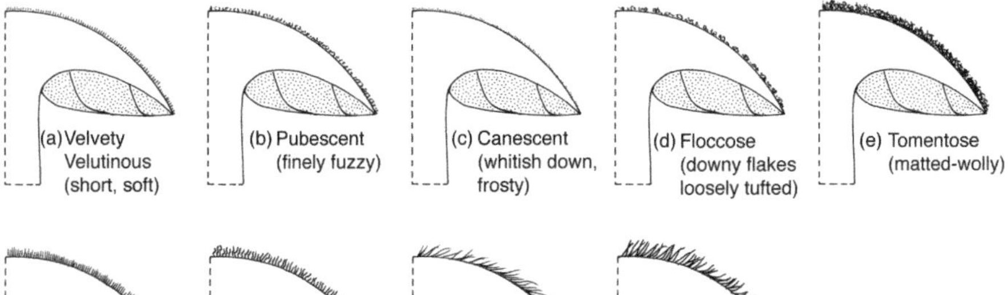

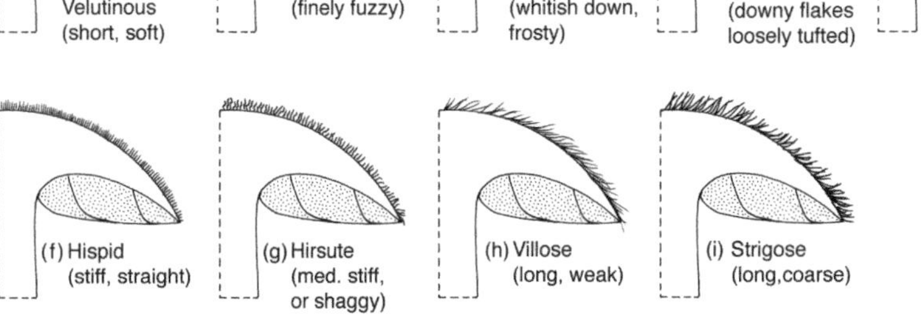

Figure 30a–i. Basidiome pileus surface hairiness morphology

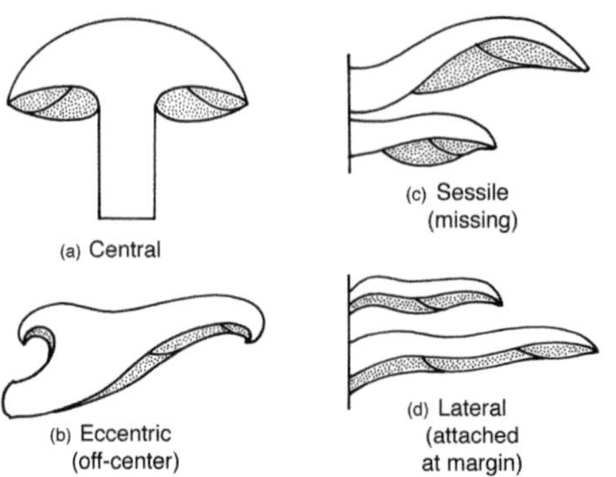

Figure 31a–d. Stipe (stalk) morphology: placement

Figure 32a–n. Stipe (stalk) morphology: shape and base

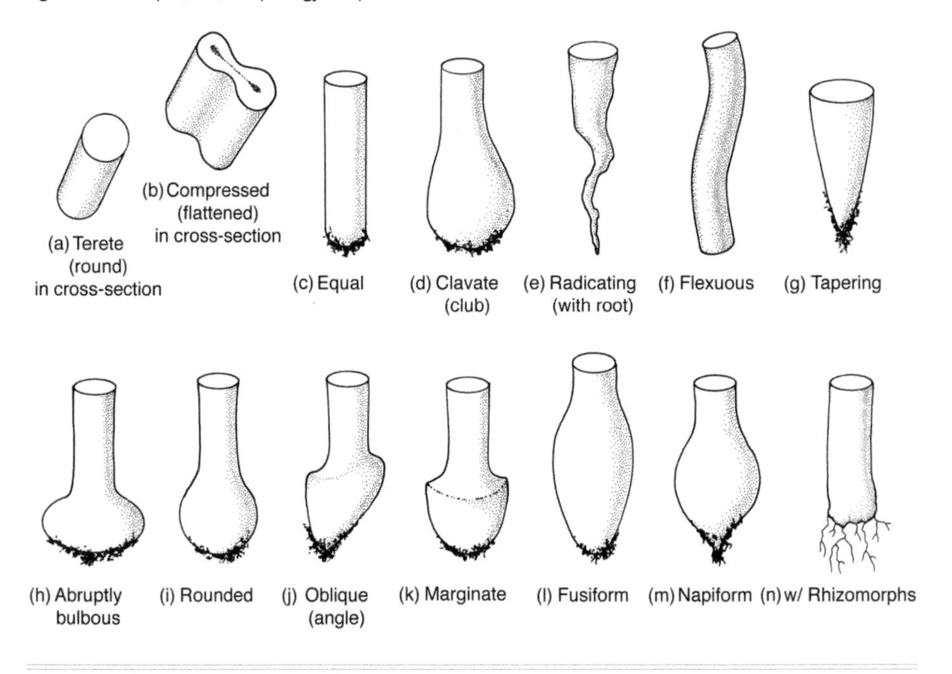

Figure 33a–g. Stipe (stalk) morphology: volva

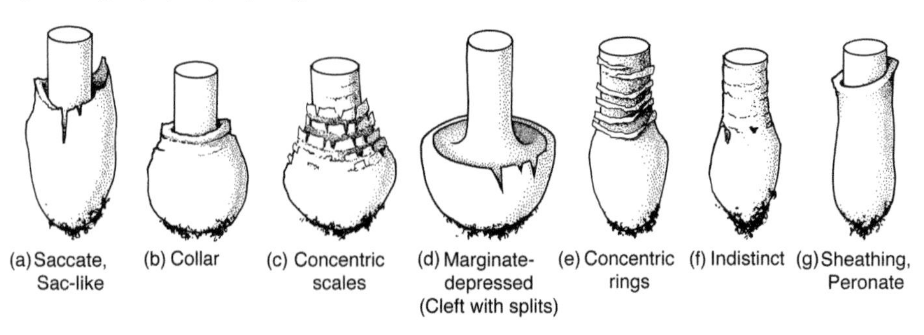

Figure 34a–l. Stipe (stalk) morphology: surface ornamentation

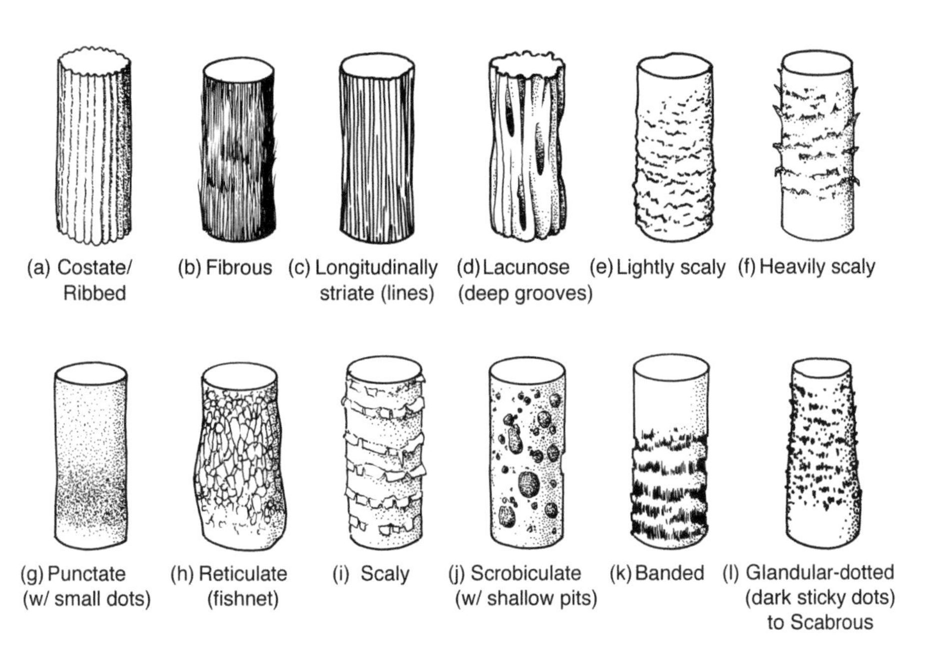

Figure 35a–f. Stipe (stalk) morphology: interior

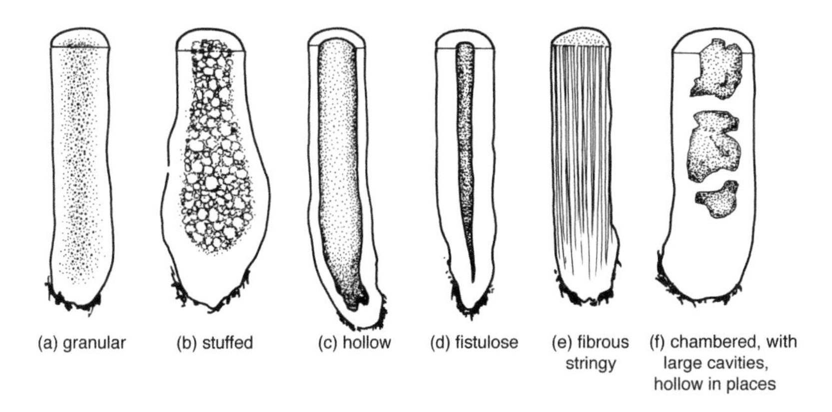

Figure 36a–j. Stipe (stalk) morphology: partial veil remnants

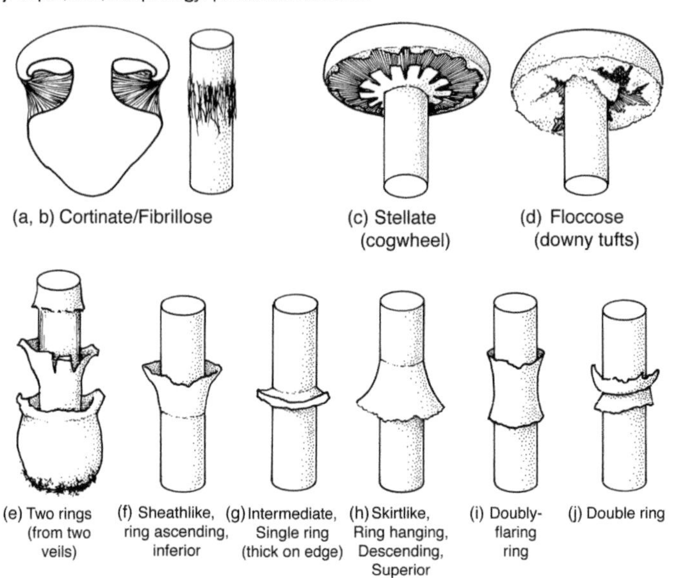

Figure 37a–o. Hymenophore morphology: spore-producing surfaces

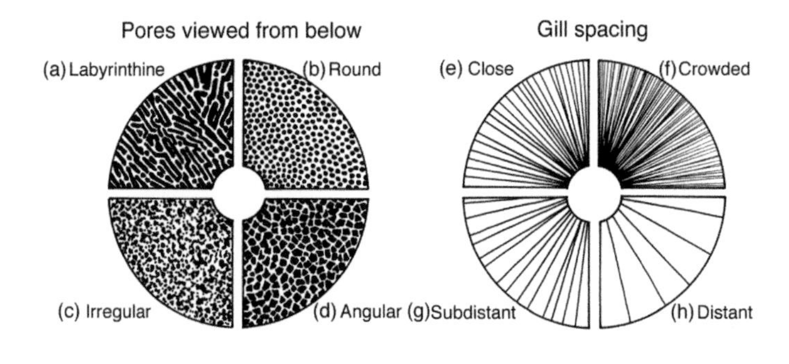

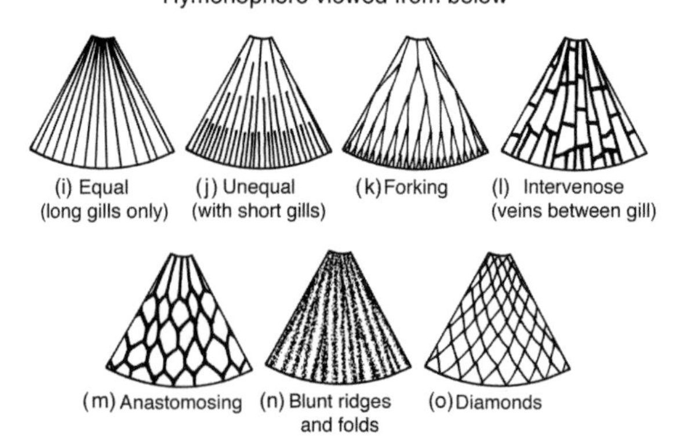

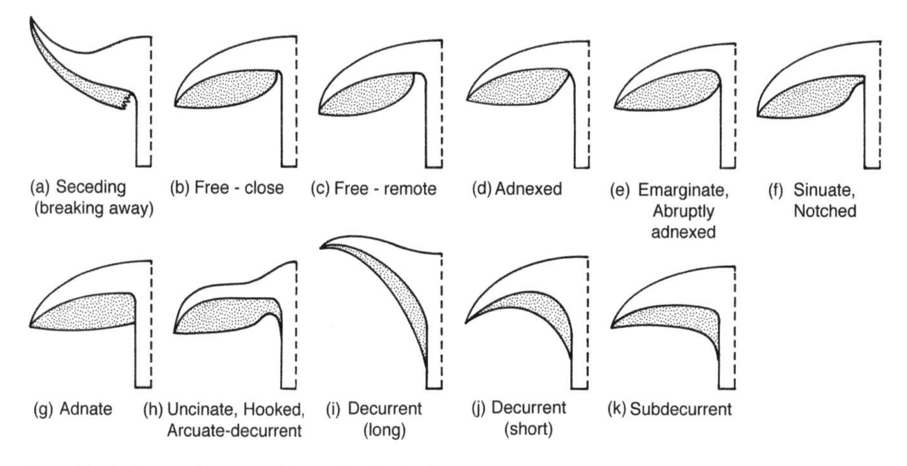

Figure 38a–k. Hymenophore morphology: lamellar (gill) attachment

Figure 39a–j. Hymenophore morphology: lamellar (gill) breadth and shape

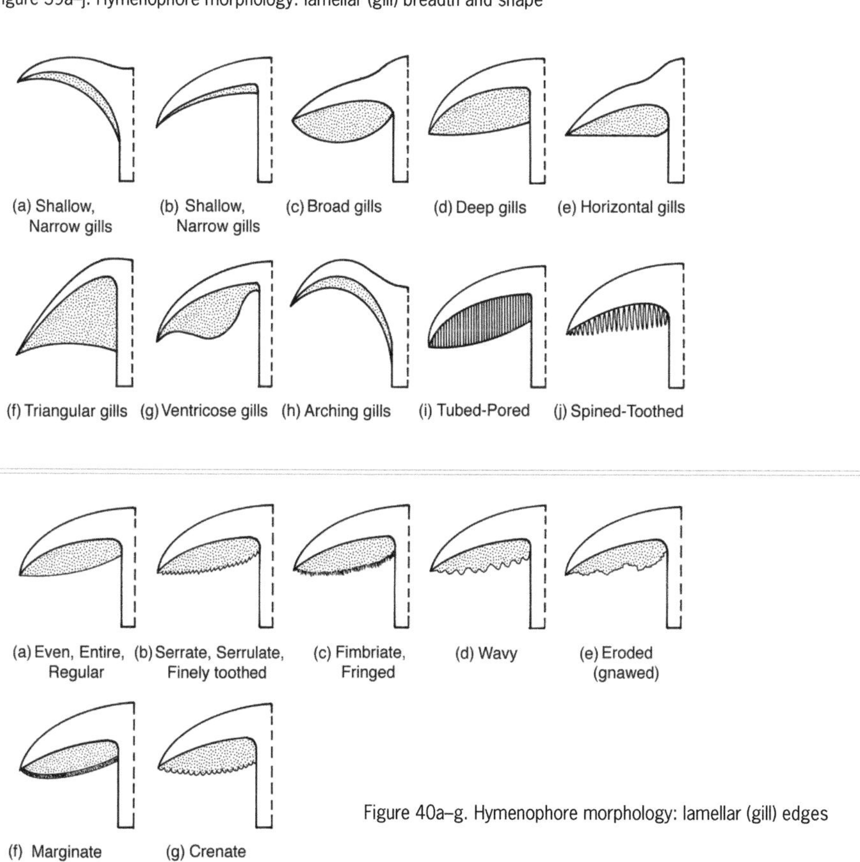

Figure 40a–g. Hymenophore morphology: lamellar (gill) edges

Figure 42. Fungal rhizomorphs

substrates (Fig. 42). Hundreds of individual and/or interconnected hyphae are the microscopic threads that represent the "feeding-by-absorption machine" of the fungus body (thallus).

Fungus or mushroom? What is the difference? "Fungus" is a broader term than "mushroom" and includes molds, aquatic fungi, the parasitic fungi that cause athlete's foot and ringworm, and other similar organisms, as well as those larger forms more easily seen with the naked eye. There is no standard architecture for a mushroom. Defined in a broad sense, mushrooms may have spore-bearing surfaces organized into gills, teeth, pores, upright branches (like a marine coral), or sacs, on and within which they produce tiny delicate spores (see Fig. 16a–j). Nor is there a mycological distinction between "mushroom" and "toadstool."

Typical gilled mushrooms (agarics) demonstrate significant variation in size, shape, and color; color changes when touched, bruised, or exposed to the air; spore color and surface textures; moisture content; interior tissues; and parts that are persistent or often disappear from obvious view with time. They may also demonstrate variations relative to their **habit**, or growth form (e.g., single, solitary, scattered, grouped [**gregarious**], and joined [**caespitose**], clustered but not joined, fused, overlapping, or in rings); **habitat**, or where they grow (e.g., on bare ground, wood, duff or humus, dung, grassy areas, forests, mosses, or plant parts); **plant associations** (e.g., trees, shrubs, herbs, algae, or cyanobaceria [blue-green algae]); and **location** (e.g., riverbeds, deciduous or coniferous [Boreal] forests, fens, marshes, bogs, shrub tundra, or alpine meadow tundra).

So much variation exists, in fact, that it is often mind-boggling, although also very intriguing to cryptogamists. Everywhere you turn, something different can be seen. Get down on your hands and knees! Let your eyes adjust. Tiny fungi you never knew existed before can be seen and represent a whole new world. Search, gather, and have fun, but **eat nothing unless you are absolutely sure the fungus you have gathered is truly edible**. The best way to ascertain this is to have it checked by someone who really knows fungi—an agaricologist (a mycologist who studies gilled fungi). Then, if "edible" as stated in current mycological literature sources, eat only a very small portion the first and second days, skip a day, and eat again on the fourth day if no ill effects are experienced. Some **toxins** are cumulative, and others are extracted if alcohol is consumed simultaneously with the fungal tissue, or are present if the fungus is not cooked. Delayed-action toxins can show no toxic symptoms for up to 24 hours to two weeks after ingestion. The recommendation is that you should always cook the fungi destined for your table.

A typical mushroom demonstrates variation in its cap (**pileus**), most often shaped like the cap of an umbrella and precariously perched on a stalk (**stipe**) (see Fig. 19).

On the underside of the cap are the spore-producing surfaces, which are organized into structures supporting the developing hymenium (gill-like plates: **lamellae**, teeth, pores, etc.). These structures are, in turn, lined by a special cell layer (**hymenium**) that contains specialized microscopic cells (**basidia**) that produce spores and/or sterile structures (**cystidia**) that act as metabolic "garbage cans" or hold the gill plates themselves apart so that spores may be "shot" off their perches to fall into air currents to be carried away, often for great distances (hundreds if not thousands of miles) by soft breezes.

When these microscopic spores land on suitable substrates or sites they may germinate, a process that results in the production of the microscopic hyphae. As previously noted, hyphae en masse form the mycelium or spawn that permeates the substrate (material on and in which the fungal hyphae grow) and gather up the needed moisture and nutrients required to maintain an assimilating livelihood. As the season progresses, mycelial masses develop and differentiate into specific tissues. With the addition of water, these tissues swell and the structure we know so well as the mushroom fruitbodies seem to literally "pop" up out of the ground, from an old rotten log, wood chips, or from plant or animal parts (dead or alive).

Figure 43. Mushroom spore print/color on white paper

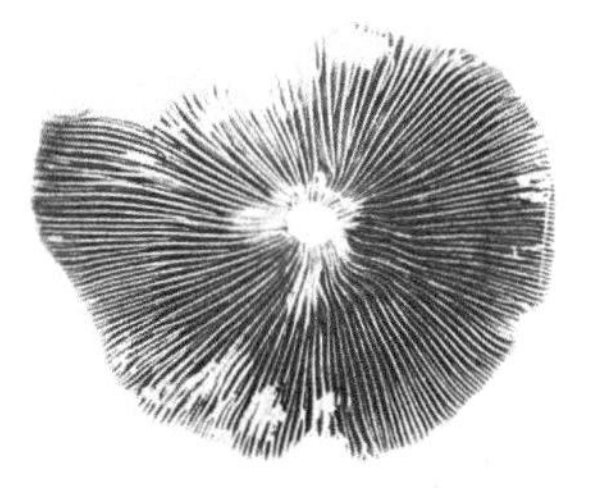

The identification of fungi in the field can be difficult without knowing salient features demonstrated by a particular species of fungus. It is important to the mushroom seeker to find the many, if not all, possible stages and conditions of the mushroom being picked for examination. One important characteristic of a mushroom is spore color (as a function of being discharged and dropped en masse onto a piece of white paper) (Fig. 43).

The spore print allows you to determine spore color, which is frequently used as the first characteristic in the identification of specimens (representing a given **taxon**). You can easily preserve the beautiful and delicate spore prints for aesthetic purposes by spraying them lightly with an acrylic or clear lacquer coating.

How to Make a Spore Print

- Clip pileus (cap) from its stipe (if present).
- Place pileus (cap) onto and over white paper, hymenium down, as often found in nature.
- Place pileus and paper into a paper container (brown lunch bag, waxed paper) or place a paper/plastic/glass cup or bowl over fungus cap on paper.
- Leave overnight in a cool place (needs the moisture for spore discharge).
- Examine by lifting pileus off paper very gently!
- Voilà! A spore print. Now let it air dry and preserve it.

Mushroom Edibility and Poisoning

Connoisseurs of wild mushrooms should take particular caution in any attempt to add new species to their repertoire of edibles. The only way to determine whether or not a mushroom is edible is to **know** the mushroom. Check the literature and field guides or consult experts, and only then, if it is known to be edible, eat a small amount with extreme caution. Your body chemistry differs from someone else's. One person may eat an "edible" mushroom with no ill effects. You, on the other hand, might develop a headache, become nauseated, downright sick, possibly losing your meal or even worse. Mushrooms of dubious identification should **never** be eaten. And, for goodness' sake, do not mix your collections when collecting. Keep your collections of fungi separated, as mixtures make it very difficult to identify the mushrooms from stomach contents.

- **Do not take unnecessary chances or make costly mistakes that may have various degrees of lasting effects (liver damage, kidney failure, death)!**
- **Do not drink alcohol with any mushroom being eaten for the first time. Some toxins are "extracted" by the alcohol and resulting illness may cause you to lose your whole dinner—several times!**

We have many "bad actors" in the Alaska interior. And do not forget, even those fungi known to be edible must be collected while young and eaten in a fresh state or condition, lest you become sickened by secondary metabolites and toxins produced by unrelated but other organisms associated with fungi, such as bacteria, fly and gnat larvae, or beetles. Do not store any mushrooms for the table in plastic bags, even if refrigerated. Even commercial mushrooms can go "bad" by doing this. This is because bacterial growth is most probable under these conditions. And then there are those fungi, like *Coprinus atramentarius* and *Leccinum atrostipitatum*, that if eaten and chased with alcoholic libations will release ipecac-like toxins and you will lose all that you have eaten, maybe even violently, and several times for hours!

Where to Look for Fungi

Each fungus species will usually be found in a specific habitat predicated on its nutritional and moisture needs. So, look on live, standing dead, and fallen trees (**lignicolous** fungi). Look in all forest types over the ground, on leaves, needles, stems, twigs, and soil (**epiphytic** fungi). Look under logs, around stumps and tree bases, under and on carpets of mosses (**muscicolous** fungi), and on old dung of moose, caribou, bear, or ptarmigan (**coprophilous** fungi). You will find many forms along river and stream courses (called **riparian** areas), in old fire burns and around campfire pits (**pyrophilous** fungi), in "dwarf" forested areas, and in the tundra. You will not find many fungi in grassy or marshy areas dominated by sedges and rushes. Seasonality is also important. If you are a visitor to the Interior during late spring or early summer (May and June), you will see more cup fungi than any other form. Midsummer (July) brings a flush of agaric and bolete fungi. But fall (late July through August and into September) yields the biggest flush of macrofungal species, some of which often occur in massive fruitings (*Agrocybe praecox*) brought by copious and welcome August rains (Fig. 44).

Each plant community has its own corresponding fungal community. You can more easily find a fungus, or identify one you have already found, if you know the plant community types in which the fungus is living. In this field guide we provide you with a macroscopic description of each species, which also includes information on the typical habitat. Descriptions of the major types of plant communities and where they are found in relation to the road are provided. Under each plant community description is a list of the species of fungi that will occur there. The habitat where a fungus is to be found (e.g., on or near specific plants or substrates in the community) is also listed in parentheses next to the name of the fungus. Some fungi are more abundant in one plant community than another and may be listed under two or more plant community types.

First, we must clarify some plant terms used in the text. Some fungi in this book are described as being associated with "ericads," or "ericaceous" understory plants. These terms refer to plants in the heath family, or Ericaceae, which are regularly associated with mycorrhizal fungi. A number of ericads are common in forest and tundra communities and include low evergreen shrubs such as mountain cranberry

Figure 44. *Agrocybe praecox* flush abundant fruiting

or lingonberry (*Vaccinium vitis-idaea*), the Labrador teas (*Ledum groenlandicum* and *L. decumbens*), heather (*Cassiope tetragona*), alpine-azalea (*Loiseleuria procumbens*), and deciduous shrubs such as bog blueberry (*Vaccinium uliginosum*) and bearberry (*Arctostphylos rubra* and *A. alpina*). Other fungi are described as being associated with dwarf birch, a term that refers to two species of shrubby birch, *Betula glandulosa* (sometimes called resin birch) and *B. nana*, which can be difficult to tell apart from *B. glandulosa*, although it generally has smaller leaves.

FOREST FUNGI—Forest communities in the Alaska interior (Fig. 45) have seven dominant tree species: white spruce (*Picea glauca*), black spruce (*Picea mariana*), quaking aspen (*Populus tremuloides*), balsam poplar (*Populus balsamifera*), black cottonwood (*Populus trichocarpa*), paper birch (*Betula papyrifera* or *B. neoalaskana*), and larch (*Larix laricina*). Upland, better-drained sites are dominated by white spruce, aspen, and paper birch. In forests along streams and rivers, balsam poplar and cottonwood take the place of the aspen, but willows (*Salix* species) are also present along with alder (*Alnus crispa*). The coldest, wettest sites are mostly black spruce, with some larch. Tall shrubs in the forest are American green alder (*Alnus crispa*) and several willows (*Salix* spp.). Low-growing shrubs include dwarf birch (*Betula glandulosa* and *B. nana*), prickly rose (*Rosa acicularis*), bog blueberry (*Vaccinium uliginosum*), mountain cranberry (*Vaccinium vitis-idaea*), Labrador tea (*Ledum groenlandicum* and *L. decumbens*), crowberry (*Empetrum nigrum*), and bearberry (*Arctostphylos rubra*).

Along the park road in the area between the park entrance and headquarters, there is a mixed forest of white and black spruce, aspen, poplar, and paper birch. The paper birch (and its specific fungi) reaches its elevational limit at this point and is not found along the rest of the park road. Aspen and black spruce (and their specific fungi) reach their elevational limits as forest trees at about the nine-mile checkpoint along the road, and are not common again until you reach the end of the road at

Figure 45. Forest communities in Interior Alaska

Kantishna. The scattered forest stands from nine-mile to Kantishna are dominated by white spruce, with some balsam poplar. Fungi found in these forest communities are listed below.

Species of Higher Fungi	***Substrates Found Near or On***
Amanita muscaria v. *muscaria*	Paper birch, aspen
A. muscaria v. *regalis*	Aspen, paper birch, occasionally spruce
A. vaginata v. *nivalis* (= *Amanitopsis hyperborea*)	Aspen, paper birch, white spruce
Hygrophorous olivaceoalbus	White spruce with ericads
Lactarius deliciosus	White spruce
L. repraesentaneus	White spruce
Russula aeruginea	White spruce with aspen
R. emetica	Spruce with hardwoods
R. subfoetens	White spruce, paper birch, aspen, blueberry
Armillaria mellea	Aspen, balsam poplar
Collybia dryophila	White spruce with aspen, paper birch, willow
Cystoderma amianthinum	Moss under closed white spruce
C. fallax	Moss under white spruce
Laccaria laccata	Moss under white spruce
Microcollybia tuberosa	Decaying mushrooms
Mycena pura	White spruce with aspen
Phyllatopsis nidulans	Dead white spruce, alder
Pluteus cervinus	Paper birch logs
Paxillus involutus	Paper birch with ericads
Cortinarius alboviolaceus	White spruce, with paper birch or aspen
C. croceus	White spruce, with paper birch or aspen
C. trivialis	Spruce with paper birch, aspen
Crepidotus mollus	Dead alder, willow, balsam poplar branches
Galerina spp.	*Sphagnum* mosses
Inocybe geophylla	Spruce
Pholiota aurivella	Dead aspen
P. squarrosoides	Paper birch
Hypholoma udum	Mosses under white spruce, aspen, and paper birch
Coprinus atramentarius	White spruce with aspen
C. micaceus	Spruce with alder, paper birch, aspen
Boletus edulis	Aspen, paper birch
Leccinum aurantiacum	Aspen
L. insigne	Aspen
L. rotundifoliae	Dwarf birch, wet places
L. scabrum	Paper or dwarf birch
Suillus cavipes	Larch, western part of park only
Plicatura nivea	Dead alder twigs

Species of Higher Fungi	*Substrates Found Near or On*
Plicaturopsis crispa	Dead alder, paper birch, willow branches
Gloeopeniophora aurantiaca	Alder branches
P. rufa	Dead aspen branches
Ganoderma applanatum	Aspen and other hardwoods
Inonotus tomentosus	Apruce
Phellinus igniarius	Hardwoods
P. pini	Spruce
P. tremulae	Aspen
Albatrellus ovinus	Spruce
Bjerkandera adusta	Hardwood logs
Cerrena unicolor	Dead hardwoods
Fomes fomentarius	Paper birch, alder
Fomitopsis pinicola	Spruce
Ischnoderma resinosum	Dead spruce, hardwoods
Piptoporus betulinus	Dead paper birch
Polyporus alveolaris	Dead hardwood branches
P. brumalis	Dead hardwood
Schizopora paradoxa	Dead hardwood branches
Trametes suaveolens	Willow and other hardwoods
Trichaptum biforme	Hardwoods
Hydnellum suaveolens	Spruce
H. imbricatum	Spruce
Clavariadelphus ligula	Spruce, hardwoods
C. pistolaris	Spruce, hardwoods
Ramaria formosa	White spruce, hardwoods
R. aff. rasilospora	Spruce, hardwoods
Thelephora terrestris	Spruce
Lycoperdon perlatum	Disturbed areas in white spruce
Nidularia aff. farcta	Brown and crumbly dead white spruce
Chrysomyxa ledicola	Foliage rust on spruce, spots on Labrador tea
C. arctostaphyli	Witches'-broom on spruce, spots on bearberry
Tremella mesenterica	Small dead branches, especially alder
Rhytisma salicinum var. arcticum	Dwarf birch leaves
Cudonia circinans	Deep mosses
Spathularia flavida	Mosses under spruce
Slime Molds	
Badhamia utricularis	Bark of decaying logs
Lycogola epidendrum	Decaying wood
Craterium leucocephalum	Dead leaves and twigs
Mucilago crustacea	Low-growing living plants, litter
Fuligo septica	Decaying wood and bark, litter and soil
Trichia varia	Stumps and logs

Species of Higher Fungi	*Substrates Found Near or On*
Lichenized Fungi	
Icmadophila ericetorum	Spruce logs, humus, mosses
Peltigera aphthosa	Moss-covered soil or decaying wood
Cladonia rangiferina	Humic soils
C. bellidiflora	Soil or humus
C. deformis	Humus or rotting wood
Multiclavula mucida	Wet rotting wood
Lichenomphalia hudsoniana	Rotten logs in shady sites

TREELINE FUNGI—Treeline plant communities in the Alaska interior (Fig. 46) are characterized by open stands of white spruce with dwarf birch and willow dominating the spaces between trees. Bog blueberry, mountain cranberry, and crowberry are abundant. There may also be grassy meadow openings with showy flowers. Along the park road, tree-line plant communities are found from five-mile to Igloo Canyon, at Toklat, and again at Wonder Lake.

Species of Higher Fungi	*Substrates Found Near or On*
Hygrophorus eburneus	With the lichen *Stereocaulon*
Phyllatopsis nidulans	Dead white spruce, alder
Cortinarius aff. *mucosus*	Reindeer lichens and *Dryas*
Crepidotus mollus	Dead alder and willow branches
Rozites caperata	White spruce and dwarf birch
Xerocomus subtomentosus	Dwarf birch and ericads
Leccinum rotundifoliae	Dwarf birch, wet places
L. scabrum	Paper and dwarf birch
Plicatura nivea	Dead alder twigs

Figure 46. Treeline plant communities

Species of Higher Fungi	*Substrates Found Near or On*
Plicaturopsis crispa	Dead alder, birch, and willow branches
Inonotus tomentosus	Spruce
Phellinus pini	Spruce
Albotrellus ovinus	Spruce
Fomes fomentarius	Birch, alder
Fomitopsis pinicola	Spruce
Hydnellum suaveolens	Spruce
Sarcodon imbricatus	Spruce
Hericium coralloides	Birch logs and stumps
Clavicorona pyxidata	Balsam poplar logs
Thelephora terrestris	Spruce
Nidularia aff. *farcta*	Brown and crumbly dead white spruce
Chrysomyxa ledicola	Spruce foliage rust, spots on Labrador tea
C. arctostaphyli	Witches'-broom on spruce, spots on bearberry
Tremella mesenterica	Small dead branches, especially alder
Rhytisma salicinum var. *arcticum*	Dwarf birch leaves
Cudonia circinans	Deep mosses
Spathularia flavida	Mosses under spruce
Lichenized Fungi	
Nephroma expallidum	Mosses
Peltigera aphthosa	Moss-covered soil or decaying wood
Cladonia rangiferina	Humic soils
C. stellaris	Soil and humus
Stereocaulon tomentosum	Open disturbed areas

TUNDRA FUNGI—Common tundra plant communities (Fig. 47) include those dominated by ericaceous shrubs, including bog blueberry, mountain cranberry, crowberry, bearberry, heather (*Cassiope*), and Labrador tea; those dominated by dryads (*Dryas octopetala* and *D. integrifolia*); and those dominated by dwarf birch with willows.

Species of Higher Fungi	*Substrates Found Near or On*
Lactarius aspideus	Dwarf willows and ericads
L. maculatus	Dwarf willows
L. rufus	Dwarf birch, well drained
Russula nana	Dwarf willows and alpine bearberry
Cantharellula umbonata	Mosses
Melanoleuca melaleuca	Mosses, ericads, dwarf willow
Paxillus involutus	Dwarf willows, ericads, mosses
Xerocomus subtomentosus	Dwarf birch
Leccinum rotundifoliae	Dwarf birch and wet places

Figure 47. Tundra plant communities

Species of Higher Fungi	*Substrates Found Near or On*
L. scabrum	Dwarf birch
Calvatia booniana	Herbaceous tundra meadows
Helvella corium	Dwarf willows
Rhytisma salicinum var. arcticum	Dwarf birch leaves
Slime Molds	
Mucilago crustacea	Low-growing living plants and litter
Lichenized Fungi	
Nephroma expallidum	Mosses and ericads
Peltigera aphthosa	Moss-covered soil or decaying wood
Umbilicaria hyperborea	Rock
Flavocetraria nivalis	Soil
Masonhalea richardsonii	In depressions, unattached
Dactylina arctica	Humic soil often with *Dryas* and ericads
Lichenomphalia alpina	Dark humus around granite rock outcrops

RIPARIAN FUNGI—Riparian plant communities grow on floodplains of rivers and streams (Fig. 48). In interior Alaska, these plant communities are similar over an elevational range from boreal forest to tundra. Riparian plant communities are dominated by tall shrubs, including willows, primarily felt leaf willow (*Salix alaxensis*) and American green alder (*Alnus crispa*). Riparian plant communities are found all along roads and wherever roads cross rivers or streams.

Species of Higher Fungi	*Substrates Found Near or On*
Armillaria mellea	Aspen, balsam poplar
Phyllatopsis nidulans	Dead white spruce, alder
Crepidotus mollus	Dead alder, willow, balsam poplar branches
Plicatura nivea	Dead alder twigs
Plicaturopsis crispa	Dead alder, paper birch, and willow branches
Peniophora auranitiaca	Alder branches
Fomes fomentarius	Paper birch and alder
Trametes suaveolens	Willow and other hardwoods
Clavicorona pyxidata	Balsam poplar logs
Arrhenia lobata	Mosses along creek and stream courses
Clavaria aff. *bovista*	Grassy river terraces
Tremella mesenterica	Small dead branches, especially alder
Mitrula borealis	Mosses and liverworts
Lichenized Fungi	
Peltigera aphthosa	Moss-covered soil or decaying wood
P. polydactyla	Soil, humus, or mosses
Stereocaulon tomentosum	Old, dry gravel bars

DISTURBED AREA AND OTHER SPECIALIZED HABITAT FUNGI—These fungi include those found on dung, burn sites, bone, hair or dead animal sites, insects, mosses, or generally "disturbed" sites.

Figure 48. Riparian plant communities grow on floodplains of rivers and streams

Species of Higher Fungi	*Substrates Found Near or On*
Hygrocybe conica	Grassy or mossy disturbed sites
Coprinus comatus	Hard ground and grassy areas
Lycoperdon perlatum	Disturbed areas in white spruce
Pholiota carbonaria	Recent burns, with *Marchantia* liverworts
Geopyxis carbonaria	Recent burns, charred spruce needles
Gyromitra infula	Recent burns, charred spruce needles
Coprinus patouillardii	Moose dung
Panaeolus acuminatus	Moose and vole dung
Hypomyces chrysospermum	Bolete mushrooms
Stropharia semiglobata	Old bear or moose dung
Lichenized Fungi	
Multiclavula mucida	Wet disturbed soil (roadsides)

Fungal Grouping (Taxonomy)

All fungi reproduce by and are dispersed as microscopic propagules called spores. Spores are produced in such numbers that when a mushroom cap is placed on a piece of white paper and left for a few hours, the forcibly discharged spores literally pile up. The pile is called a spore print. Depending on the fungus, the spore print will have a color (as viewed en masse) ranging from white, pastel, pinkish, yellow, rusty, brown, purple brown to black. The configuration of the spore-bearing surface (**hymenium**), from whence these tiny spores come and on which they are produced, is important for determining the identity of a mushroom. Their spore-bearing surface may also vary from smooth, toothed, to gilled, depending on the type of fungus.

Fungi you are most likely to encounter fall into one of many commonly seen and collected taxonomic groups that differ in spore color and the basic arrangement of the spore-bearing surface, such as on gills (lamellae), pendent teeth, upright coral-like branches, wrinkles and ridges, in spongy pores, corky pores, and cups, etc. These groups are commonly referred to as agarics, boletes, bracket fungi or polypores, tooth fungi, coral fungi, chanterelles, puffballs, thelephores, earth tongues, cup fungi, and slime molds, respectively. Fungi also may be partitioned into ecological groups (i.e., as a function of the role they play in interior Alaska settings). They include the many saprobes or saprotrophs (decomposers of dead plant and animal tissue), parasites, mycorrhizae (fungus-root relationships), lichenicolous fungi (saprophytes and parasites of lichens), and the lichenized fungi (lichens) that are often early colonizers of new or bare substrates. They are all quite interesting, but the real treat is in searching for and collecting particular examples of the myriad forms that frequent the Alaska interior.

To begin identifying a fungus, there are a number of important questions to ask and features to look for. Major questions to ask are: How is the spore-bearing surface organized? What form does it take? Does it have gills, pores, pendent teeth, wrinkled surfaces, upright clubs, or blunt ridges? Is the hymenium exposed or "enclosed"?

The table below will assist your placement of fungi found into some sort of common grouping. With a spore print (spore color), you will be able to go to many of the field guides for fungi from other regions, most of which assume you have an inkling of fungal families, and begin the keying process to put a name on the collection. The following table shows fungal groups by hymenial organization:

Fungus Group	*Common Name*	*Hymenium Organization*
Agaricoid fungi	Gilled fungi	Lamellae or gills
Boletinoid fungi	Boletes, fleshy pored fungi	Spongy tubes with pores that peel off as a thick mass
Polyporoid fungi	Pored or tubed woody, corky Leathery fungi	As crusts, corky brackets, tubed or pored shelves
Hydnoid fungi	Spine or hedgehog fungi	As pendent teeth under cap
Coraloid fungi	Club fungi	As single to branching, upright clubs
Cantharelloid fungi	Chanterelle fungi	As blunt ridges looking like precursors to "gills"
Thelephoroid fungi	Vace-like fungi	As smooth, bumpy, or winkled surfaces
Gasteroid fungi	Stomach fungi	
Puffballs		As "stomach" balls of dry brownish powder when mature
Bird's nests		As small flattened "eggs" in a vase-like, sessile "bird's nest" on woody substrates
Rusts	Conifer rust fungi	As brightly colored (yellow-orange) pustules
Jellies	Jelly fungi	As wood-inhabiting, brightly colored gelatinous masses
Cups	Sac fungi	As inner surfaces of tan to brightly colored "cups" that occur close to the substrate
Earth tongues	Sac fungi	As little "spades" sticking up and out of the mosses
Slime molds	Plasmodial "fungi"	As minute sessile or stalked balls, plumes, tubes, and other shapes more difficult to characterize
Lichens	Lichenized Sac and Club fungi	Fungal (mycobiont) Algae (phycobionts) relationships
Ascolichens Sac fungi	Crustose	As oddly colored black to yellow to bluish-green crusts
	Foliose	Leaf-like
	Fruticose	Tree-like forms attached to a variety of substrates
	Squamulose	A mixture of foliose and fruiticose forms
Basidiolichens	Coral	Upright, +/– branched
	Agaric	Gilled
Club fungi	Agaric	Gilled
	Boletes	Pored, fleshy
	Bracket	Pored, tough
	Coral	Upright, branched

Tools of the Trade

No vocational handbook or guidebook would be complete without its list of useful items. Some high-tech enthusiasts come loaded to the gills, no pun intended! But there are some things you just cannot do without. The equipment listed below will be useful whenever pursuing the elusive mushroom, whether purely for its beauty, varying shapes and forms, or for the table. The following table shows useful items to mushroom collectors:

Item	*Use and Utility*
Clothing:	
Boots	Rubber boots or hiking shoes for tromping through the woods and over tundra
Rain gear	Pants, jacket with hood and pockets, wool socks
Layers	Polypropylene, wool (little or no cotton)
Hat	Preferably with a sun shade
Gear:	
Notepad and pencil	For taking notes on such characteristics as color, texture, odor, and, in some cases, taste
Hand lens (10x to 20x)	For magnifying the hard-to-observe spore-bearing surfaces while in the field, as well as other parts of the mushroom
Compass	For trekking in and out of the woods
Camera/film	35mm with a macro lens, flash, 25–64 or 100–200 ASA film for capturing the true beauty of the fungi in their natural settings
Digital camera	Memory cards to hold lots of pictures at reasonably high resolution, a macro lens, and a battery charger!
Whistle/bell	For alerting bears to your presence or informing your collecting partners of your whereabouts
Knife	Pocket knife or, preferably, sheath knife for getting the "whole thing," remembering that it will be difficult to impossible to key out your fungus to its proper group or name without all parts being available for examination
Gruber rake	For truffle hunting
Hatchet or axe	Small, for wood substrate fungi
Twig snips	For cutting twigs with fungi on them
Saw	For sawing larger branches
GPS	Global/geographical/ground position system for recording collection localities
Tape recorder	For making verbal notes on important bruising reactions, exudate, and/or tissue oxidation color changes
Collecting container and contents:	
Basket	With body so fungi are not crushed (with rigid sides; see Fig. 49)
Waxed paper	Nonplastic paper sheets or sandwich bags work well for most specimens; waxed paper sheets will suffice (plastic bags do work well for mosses and lichens)
Fungus labels	For numbering, making notes, and spore printing
Paper bags	For sorting the myriad fungi found
Containers	Small plastic, cardboard, or metal for tiny, fragile fungi
Bug dope	Of the strongest kind (high percent of active ingredient) or a strong will!

(continued on next page)

Item	*Use and Utility*
Food and drink:	
Water	For drinking
Cheese	For munching
Crackers	For munching
The unusual:	
Meteorological station	To capture daily recorded wind speed, direction, temperature, relative humidity, and precipitation (Fig. 50)
Fungus dryer	Preserving labeled treasures for storage and safe keeping (Figs. 51 and 52)
Enthusiasm:	
Lots of it	For probing the little-known world of the mystical, but not so magical, mushrooms
Cautions:	
Firearms	Unless absolutely necessary to protect against megafauna, leave them at home. They are illegal to have in some places (DNP&P)
Safety	Always an important aspect in collecting mushrooms, particularly in Alaska

Figure 49. Collecting basket

Figure 50. Meteorological station

Figures 51 and 52. Fungus dryer: specimen preservation

Fungal Groups: The Hymenomycetes

(Having an Organized Hymenium)

Agaricoid (Gilled) Fungi

To truly understand this diverse group of fungi, it is necessary to consider the surface (hymenium) upon which minute spores are produced. Agaric (from "Agaria," a city in Sarmatia of ancient Greece) refers to the plate-like gills, the spore-bearing structures found hanging beneath the cap in this group of fungi (see Fig. 16a). Gill plates of varying thickness, length, width, and separation distance (especially the knife blade–like edge) make up what is known as the **hymenophore**. It is on those flat gill "faces" that spores are borne and forcibly discharged from the tips of specialized cells called basidia. The spores develop on tiny, specialized hornlike protrusions called **sterigmata**. All members of this group possess gills on the underside of their caps (**pileus**). A beautiful spore print can often be made by removing the stalk (**stipe**) from the cap where these are joined and placing the cap, gill side down, over a piece of white paper and covering for several hours. (Unfortunately, this is not the case with gastroid forms of agaric fungi that have "gills" but lack forcible spore discharge; hence, no spore print is possible.) A cup or bowl should be put over the cap to retain the moisture needed by the fungus for forcible spore discharge. In some species of mushrooms, there is a "skin" (veil) that will cover the entire mushroom when the button is very young. A complete covering is termed "universal" (**universal veil**). Alternatively, mushrooms may display a skinlike or cobweblike tissue that covers only the hymenium (**partial veil**) and may be attached to the stalk and/or to the margin of the cap. This latter tissue may become detached from the cap margin early in development and collapse down around the stalk, where it may even disappear completely in due time! If the partial veil is retained as a membranelike ring or a weblike cortina, it is referred to as an **annulus** or **veil**.

Representatives of seventeen taxonomic **families** of agarics (gilled mushrooms) are commonly found in Interior Alaska. Agarics you are most likely to find will probably belong to one of fifteen key families: Amanitaceae, Hygrophoraceae, Lepiotaceae, Marasmiaceae, Pleurotaceae, Tricholomataceae, Russulaceae, Pluteaceae, Entolomataceae, Bolbitiaceae, Paxillaceae, Cortinariaceae, Strophariaceae, Psathyrellaceae, or the Coprinaceae. There are several dozen **genera** of agaric fungi found in the Alaska interior, but only twenty-five are herein considered.

Edibility has not been established for many of the Interior's agaric fungi. Some are delicious; others are decidedly poisonous. Those that one person may eat may not be edible or digestible to another. We suspect an individual's body chemistry plays a significant role here. In any case, **do not eat any agaric** unless you know the mushroom to be edible or you know someone of authority who can attest to its edibility.

FAMILY: Pluteaceae (Amanitaceae)

GENUS AND SPECIES: *Amanita muscaria* var. *muscaria* (L.) Lam. (1783) (Figs. 53 and 54)

COMMON NAME: Fly agaric, fly (poison) amanita, or false orange

STRIKING FIELD CHARACTERS: Cap deep red, fading to orange or yellow, smooth and tacky, with white pyramidal warts (remains of a broken universal veil); white free gills; thick white stalk with a membranous ring and a bulbous base, with the upper portion of the latter breaking up into concentric rings.

MACRO AND MICRO DESCRIPTION: **Cap** 2.5–23 cm broad, narrowly convex (oval or derby-shaped), expanding to convex, broadly convex to plane in age; moist, sticky to viscid; deep red (mostly) to red-orange, orange, orange-yellow to yellow; with white pyramidal warts not easily rubbed off, dense near margin and becoming widely spaced over disc; context (flesh) white, firm, thick; margin inrolled slightly at first, becoming straight with age. **Gills** crowded, moderately thick, broad and with fuzzy edges (especially near margin), free from stalk. **Stalk** 5–20 cm long, 15–20 mm thick; expanding or straight toward a bulbous base; white; almost smooth to flocculent (tufted cottony) with a broad white, persistent skirting, hanging, membranous (partial) veil; top of bulbous base with universal veil remnants broken up into 3 to 5 concentric rings. **Spores** white in deposit, broadly elliptical, smooth, non-amyloid, 9–13 X 6.5–9 µm.

HABITAT AND ROLE: This is an ectomycorrhizal associate (terrestrial, but connected to the root system of its host) with paper birch and quaking aspen, and probably several other plant species. The fungus can be found throughout interior Alaska, singly to scattered, quite abundantly some years, and comes in no less than six varieties (see table on next page).

EDIBILITY, TASTE, AND ODOR: *Amanita muscaria* is toxic and contains three different types of toxins. It has been used recreationally for its one psychoactive component, but also may cause 12 hours of frightening toxicity for which mycologists have been called to hospital bedsides of very sick consumers. Death has been known to occur from eating this fungus. Taste is mild, and odor is pleasant. However, one should never eat any *Amanita*!

Figure 53. *Amanita muscaria* var. *muscaria* (L.) Lam. (1783)

Figure 54. *Amanita muscaria* var. *muscaria* (L.) Lam. (1783), young button stage

This mushroom, in its young button stage (see Fig. 54), resembles and may be mistaken for some edible puffballs (see figures in later section). Slice the fungus in half to make sure it is not a gilled mushroom button rather than a puffball.

Amanita muscaria

Variety	Cap Color	Universal Veil Color	Stipe Color
muscaria	Red	White	White
flavivolvata	Orange-red	Tan-yellow	White
formosa	Orange-yellow	Yellow-tan	Yellow-tan
persicina	Melon	Tan-yellow	Tan-yellow
regalis	Brown	Tan-yellow	Tan-yellow
alba	White-buff	White-tan	White-tan

FAMILY: Pluteaceae (Amanitaceae)

GENUS AND SPECIES: *Amanita muscaria* var. *regalis* (Fr.) Sacc. (1887) (Fig. 55)

COMMON NAME: Panther, panther cap, or false blusher

STRIKING FIELD CHARACTERS: Like *Amanita muscaria* var. *muscaria,* but appearing less robust; cap pale tan, gray-tan to tawny brown (no yellow, orange, or red tints); more slender stalk with "stretch mark" patches on the stalk near its bulbous and collared base.

Figure 55. *Amanita muscaria* var. *regalis* (Fr.) Sacc. (1887)

MACRO AND MICRO DESCRIPTION: **Cap** 1.5–7.5 cm broad; pale tan, grayish-tan, tan to tawny-brown; subviscid (sticky to tacky); convex, broadly convex in age; ivory-colored pyramidal universal veil remnants over cap, but thicker toward the faintly striate margin; context (inside) flesh fairly firm, ivory (just under cap) to white, thick. **Gills** free, close together, edges finely floccose (cottony or fluffy), white, and broad. **Stalk** 1.5–10 cm long, 7–15 mm wide, white and smooth near apex, surface breaking up into fibrils below annulus and appearing "stretched" (lacerate) and expanding slightly toward the bulbous but collared base. **Spores** white in deposit, broadly elliptical, smooth, non-amyloid, 9–13 X 6.5–9 μm.

HABITAT AND ROLE: This stately mushroom is terrestrial, solitary to gregarious, and is a suspected mycorrhizal associate with aspen, birch, and (occasionally) white spruce.

EDIBILITY, TASTE, AND ODOR: Deadly poisonous, this mushroom should never be eaten. Taste is mild, odor is pleasant, but reminds one of what something bitter tasting would smell like. Once again, *never* eat this or any other *Amanita.*

FAMILY: Pluteaceae (Amanitaceae)

GENUS AND SPECIES: *Amanita vaginata* f. *vaginata* (Bull.) Lam. (1783) (Fig. 56)

COMMON NAME: Grisette, ringless amanita

STRIKING FIELD CHARACTERS: Cap gray to gray-brown, tacky, smooth, with a striate margin; no annulus; with a volvate (saclike) stipe base and free gills.

Figure 56. *Amanita vaginata* f. *vaginata* (Bull.) Lam. (1783)

MACRO AND MICRO DESCRIPTION: Cap 3.5–7.5 cm broad; convex then expanding in age to broadly convex with broadly umbonate disc; smooth, sticky to tacky, gray to gray-brown; universal veil elements absent; flesh white, fragile, thin; margin distinctly striate (marked with delicate lines). **Gills** white, crowded, free, thin, deep. **Stalk** 6–10 cm long, 7–12 mm wide, white, smooth, without an annulus, tapering toward apex; base abruptly expanded, clublike, sheathed by a thin, white membranous saclike and persistent volva. **Spores** white in deposit, globose to subglobose, smooth, non-amyloid, 8–12 μm.

HABITAT AND ROLE: Terrestrial, solitary to few in a group, in mixed woods; a suspected mycorrhizal associate of aspen, birch, and white spruce.

EDIBILITY, TASTE, AND ODOR: This mushroom is described as being nonpoisonous or edible with caution. Taste (fresh) and odor are mild. However, one should *never* eat this or any other *Amanita.*

FAMILY: Tricholomataceae (Hygrophoraceae)

GENUS AND SPECIES: *Hygrocybe conica* (Scop.) P. Kumm. (1871) (Fig. 57)
= *Hygrophorus conicus* (Scop.) Fr. (1838)

COMMON NAME: Cone-shaped or conic hygrocybe (hygrophorus), witch's hat

STRIKING FIELD CHARACTERS: Bright yellow, orange-yellow, orange, to orange-red, conic cap; candy-cane (twisted) stalk with yellow-green tints that bruise purple-black; sticky cap; hollow stipe.

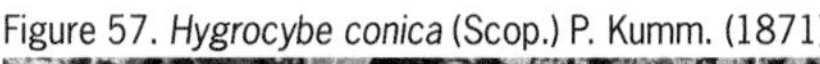

Figure 57. *Hygrocybe conica* (Scop.) P. Kumm. (1871)

MACRO AND MICRO DESCRIPTION: Cap 10–35 mm broad, conic with blunt to rounded apex, broadly expanding to form an umbo; tacky to viscid; smooth with orange to reddish orange fibrils, bruising black; flesh yellowish to yellow-orange, thin. **Gills** are close to subdistant (somewhat separated), thin, narrow, yellow in youth becoming the same color as but lighter than cap and blackening when bruised, waxy if rubbed vigorously between fingers. Gill trama is parallel (regular). **Stalk** 2.5–7.5 cm long, 3–8 mm broad, smooth, moist, candy-cane twisted with longitudinal lines, orange-yellow (olive); base is the first part of the fruiting body to show signs of bruising black; hollow. **Spores** white in deposit, elliptical, smooth, 8–14 x 5–7 µm.

HABITAT AND ROLE: Scattered to gregarious in grassy to mossy disturbed sites (old paths, runway edges, roadsides, and stream bank terraces), terrestrial.

EDIBILITY, TASTE, AND ODOR: The edibility of this fungus is questionable. It is reported as being edible or nonpoisonous yet is thought to be responsible for at least four deaths in China and may possibly even be hallucinogenic. Taste is mild and odor is pleasant. One should simply not risk eating this fungus.

FAMILY: Hygrophoraceae

GENUS AND SPECIES: *Hygrophorus eburneus* (Bull.) Fr. (1838) (Fig. 58)

COMMON NAME: Ivory hygrophorus

STRIKING FIELD CHARACTERS: Cap and stalk are white and tacky (viscid) to weakly slimy; gills decurrent.

MACRO AND MICRO DESCRIPTION: Cap 1–5 cm broad, convex, expanding to broadly convex with a subumbonate disc and inrolled smooth margin; smooth; white, to ivory to very light buff in age; flesh thick, white. **Gills** decurrent, tiered lamellulae, spaced but neither distant nor crowded, thick, waxy, white to ivory in age. **Stalk** 2.5–5 cm long, 3–7 mm wide, mostly equal (gills continuous from stipe to margin) but flexuous, pliant (easily bending) with apex breaking up into minute scales, white, lubricous (slippery) in wet weather. **Spores** white in deposit, elliptical, smooth, 6–9 x 3.5–5 µm.

Figure 58. *Hygrophorus eburneus* (Bull.) Fr. (1838)

HABITAT AND ROLE: This fungus is often found in prolific fruitings at or near treeline in association with the foliose lichen *Stereocaulon tomentosum* at the edge of the white spruce forest where the latter grades into a dwarf birch community type. It is a suspected mycorrhizal fungus.

EDIBILITY, TASTE, AND ODOR: Some reference works report *Hygrophorus eburneus* to be edible, but others consider the species of unknown edibility. At best, our Alaska form is questionable and best left off the table. Taste is mild; odor is pleasant to somewhat fragrant. This fungus is the only pure white Hygrophorus we have seen in interior Alaska, should be easily identified and not confused with Limacella having free gills.

FAMILY: Hygrophoraceae

GENUS AND SPECIES: *Hygrophorus olivaceoalbus* (Fr.) Fr. (1838) (Fig. 59)

COMMON NAME: Slimy-sheathed waxy cap, olive hygrophorus

Figure 59. *Hygrophorus olivaceoalbus* (Fr.) Fr. (1838)

STRIKING FIELD CHARACTERS: Having smoky gray, streaked, and slippery cap; stalk annulate, white above with white, distant, and decurrent gills.

MACRO AND MICRO DESCRIPTION: Cap 2.5–5 cm broad, convex, expanding to flat and finally with a depressed disc; smooth slimy in wet weather; smoky gray to almost black and with streaking fibers, darker and solid over disc, ashen (grayish brown) over inrolled margin. **Gills** subdistant, white, thickened, waxy, decurrent, and fairly broad. Gill trama divergent (irregular). **Stalk** 5–7.5 cm long, 4–7 mm wide, pure white above annuluslike ring, broadly streaked, gray below; thin blackened veil, smooth. **Spores** white in deposit, elliptical, smooth, 9–12 x 5–6 µm.

HABITAT AND ROLE: This fungus fruits under white spruce, but it is suspected of forming mycorrhizae with ericaceous understory plants such as lingonberry and crowberry. **Fruiting bodies** occur singly to scattered on moss-covered soils.

EDIBILITY, TASTE, AND ODOR: *Hygrophorus olivaceoalbus* is described by some as being "edible with caution," but unless a profuse fruiting is found, it probably would not be worth the trouble to cook up only a few fruiting bodies. Taste and odor are mild. This fungus is very recognizable and not easily confused with any others we have seen in the Alaska interior.

FAMILY: Russulaceae

GENUS AND SPECIES: *Lactarius aspideus* (Fr.) Fr. (1838) (Fig. 60)

COMMON NAME: Yellow tundra lactarius

STRIKING FIELD CHARACTERS: Small, cream to buff yellow, azonate when young to subzonate in age becoming weakly concentrically color banded near the edge when older, not banded when young, viscid (tacky) to shiny, yellow-brown silky margin and white milk changing to pink and finally lilaceous (pale purple-pink). The latex is white, but stains the flesh and gill tissues lilaceous upon oxidation.

Figure 60. *Lactarius aspideus* (Fr.) Fr. (1838)

MACRO AND MICRO DESCRIPTION: Cap 1.5–4.5 cm broad, somewhat irregular, convex to convex-depressed in age; viscid (tacky) to glutinous (slimy); light cream to buff yellow to pale yellow becoming darker with age; smooth, glabrous (not hairy) except for a butterscotch-colored, silky, and inrolled margin; flesh white, staining to lilac. **Gills** thick, fleshy, to 2 mm broad, subdistant and somewhat separated, broadly attached, and white to peach yellow. **Stalk** 1.3–2 cm long, up to 16 mm broad, smooth, dry, straw to buff yellowish, irregular, and hollow. **Spores** yellow in deposit, elliptical, reticulate, 7.5–10.5 x 7–8.5 µm.

HABITAT AND ROLE: This small but conspicuous golden yellowish mushroom of green tundra meadows is most probably a mycorrhizal associate with ericaceous plants (e.g., blueberries) and specifically with dwarf willows. It is not commonly seen. When found, it is solitary to gregarious (in groups) on south- and west-facing slopes with moist peat soil.

EDIBILITY, TASTE, AND ODOR: Edibility of this fungus is simply not known. Taste is mild to slightly acrid, and the odor is pleasant to fruity.

FAMILY: Russulaceae

GENUS AND SPECIES: *Lactarius deterrimus* (Gröger) Hesler & A.H. Sm. (1979) (Fig. 61)

COMMON NAME: Orange-latex, milky or delicious lactarius

STRIKING FIELD CHARACTERS: Large, convex to expanded, carrot orange, zoned cap and green discolorations or staining, tacky, orange latex, with a lighter carrot orange stalk and mild taste.

MACRO AND MICRO DESCRIPTION: **Cap** 2.5–10 cm convex to broadly so; smooth; disc depressed and deeply so with age; frosted whitish to pinkish orange at first becoming bright orange with multiple zones and dull orange with increased green staining with age; tacky to slippery in wet weather; flesh pale light orange to flesh-orange, staining dull greenish orange in age; margin inrolled at first, expanding in age. **Gills** thick, spaced, broad, broadly attached to very shortly decurrent; orange, more orange than cap, and staining green on edges and where bruised. **Stalk** 3–10 cm long, thick, short, 2–3 cm broad and subequal, expanding a bit at apex and base in youth (the longer it is, the more equal it becomes with a narrow blunt base in age); light pinkish orange, staining green; smooth and dry to moist, but never slimy. **Spores** pale buff in deposit, broadly elliptical, reticulate, 7.5–9 x 6–7 µm.

Figure 61. *Lactarius deterrimus* (Gröger) Hesler & A.H. Sm. (1979)

HABITAT AND ROLE: This fungus is terrestrial, scattered to gregarious in deep moss carpets under mixed white spruce, paper birch, and quaking aspen in upland forests and is a suspected mycorrhizal symbiont with understory ericoids and white spruce. It is most often confused with *Lactarius deliciosus* var. *areolatus* A.H. Sm. (1960) in our high-latitude boreal forests.

EDIBILITY, TASTE, AND ODOR: This is one of our best edibles, easily identified, and when fruiting is abundant enough for a good meal. Its taste is mild to weakly acrid (radish-like) with age. Cut your morsels for the table in half to make sure the flies have not chosen your mushrooms for the same reasons you have, lest you get a bit more protein than planned!

FAMILY: Russulaceae

GENUS AND SPECIES: *Lactarius maculatus* Peck (1891) (Fig. 62)

COMMON NAME: Grape-colored lactarius

STRIKING FIELD CHARACTERS: Large, slimy to sticky, light grayish brown to purple-brown with waterlike spots, zoned, pointed umbonate (papillate), white latex that stains gills lilac to dull violaceous (purple).

MACRO AND MICRO DESCRIPTION: **Cap** 5–7.5 cm broad, convex to broadly convex, finally convex-depressed with pointed umbo over recessed disc; color drab grayish brown to purple-brown with "water spots"; distinctly zonate; surface uneven, "smooth"; viscid to glutinous; margin inrolled at first, soon expanding; flesh white to drab, grayish-purple brown (vinaceous) beneath cuticle, staining lilac. **Gills** broadly attached to subdecurrent; whitish cream to buff, becoming dirty pale-colored to brownish, staining lavender (lilaceous); subdistant, moderately thick, to 5 mm broad. **Stalk** 2.5–4 cm long, 0.8–1.2 cm broad; "spotted," staining concolorous with (same color as) gills; hollow; subclavate (club-shaped) and tapering slightly toward base then flaring into sub-bulbous stipe base, dry to slightly sticky; pallid (pale), concolorous with gills. **Spores** yellowish white in deposit, elliptical, warted and ridged, 8–12 x 7–8 µm. Latex white to creamy-white, staining tissues purplish.

Figure 62. *Lactarius maculatus* Peck (1891)

HABITAT AND ROLE: This fungus frequents alpine meadows and is a mycorrhizal associate of willow at higher elevations. It appears close to *Lactarius cordovaensis*, which is said to be a broadleaf tree associate, and also could be quite close to *L. pseudouvidus* from European alpine zones. A lot of taxonomic research is still very much needed for these and other Alaska Lactarii.

EDIBILITY, TASTE, AND ODOR: This fungus is best left alone, since it is considered poisonous by some authors. We know very little about its edibility. Taste is mild to often bitter or with acrid (peppery) flavors, which does not render it particularly palatable. Odor is often mild to fragrant.

FAMILY: Russulaceae

GENUS AND SPECIES: *Lactarius representaneus* Britzelm. (1885) (Fig. 63)

COMMON NAME: Yellow northern spruce lactarius

STRIKING FIELD CHARACTERS: Large, dull to bright yellow, yellow-orange, densely bearded cap and margin; margin tacky to shiny, subzonate; white latex stains tissues lilac; in white spruce forests; with nutmeg odor and acrid (bitter) taste.

Figure 63. *Lactarius representaneus* Britzelm. (1885)

MACRO AND MICRO DESCRIPTION: **Cap** 2–13 cm broad, convex, convex-depressed and may become deeply depressed in age; tacky in dry weather to very slimy or gooey in wet weather; zoned with honey to butterscotch, coarse fibrils which are abundant over the heavily bearded margin; fibrous margin white at first becoming yellow to dull yellow to yellow-orange; disc often smooth; flesh white, staining lilac when exposed to air, white milk unchanging or turning lightly wheylike (watery white); flesh firm to brittle. **Gills** 5–8 mm broad, close to subdistant, broadly attached to slightly decurrent; often contorted from convex cap; light cream yellow to paler buff-yellow, spotted (especially edges) by latex when fresh. **Stalk** 2.5–9 cm long, 1.5–4 cm broad, relatively short, stocky, equal to irregular, hollow, dry to shiny, white glistening dots at apex, yellow to yellow-buff, pitted; flesh white, staining same as gills and cap tissues. **Spores** whitish to yellowish, broadly elliptical, warted and ridged, 8–12 x 6.5–9 µm. Latex white to creamy white, staining tissues lilac to dull purple.

HABITAT AND ROLE: We find this fungus almost always in white spruce to mixed white spruce, paper birch, and quaking aspen forests, projecting from the soil through thick moss carpets. *Lactarius representaneus* is most probably a mycorrhizal associate of the spruce and/or understory ericaceous (evergreen) shrubs. Most often scattered to occasionally gregarious, the tissues of this fungus stain lilac quickly in wet weather when copious white milky juice is present.

EDIBILITY, TASTE, AND ODOR: The large size, fragrant (nutmeg) odor, and mild-to-acrid (peppery) taste of this fungus might suggest something good to eat, but DO NOT eat it! Several authors report that no *Lactarius* that stains lilac or violaceous should be eaten. They can cause mild to severe gastrointestinal upset. This complex of fungi has two or three closely related species in our northern forests that need substantial work. So why take the chance of ruining a good hike?

FAMILY: Russulaceae

GENUS AND SPECIES: *Lactarius rufus* (Scop.) Fr. (1838) (Fig. 64)

COMMON NAME: Red lactarius

STRIKING FIELD CHARACTERS: With a red-brown dry cap, cream to pinkish buff short decurrent gills, and a very strong acrid (radish) bite to the flesh.

MACRO AND MICRO DESCRIPTION: **Cap** 2.5–22 cm broad, convex to convex-depressed, umbonate; dry, reddish orange to red-brown, darker over disc and often paler pinkish red over ribbed ("striate") and wavy margin; margin inrolled and silky flocculose in youth, expanding to straight and becoming broken up (lacerate) with age; flesh moderately thick, to 10 mm, light flesh tan; latex white, very acrid (peppery or bitter) and unchanging in color. **Gills** broadly attached to very short decurrent, close, 2–4 mm broad, pallid (pale), dirty cream to pinkish tan, not discoloring due to latex stains. **Stalk** 2.5–13 cm long, 5–14 mm broad, brittle, equal to slightly flexuous and slightly narrower at base, sheathed with white mycelium; dry, glabrous (hairless), smooth; pinkish flesh

Figure 64. *Lactarius rufus* (Scop.) Fr. (1838)

to rose-tinted above, otherwise the same color as the gills; stuffed to hollow. **Spores** are cream to pale yellow, elliptical, amyloid warts and ridges, 7.5–11 x 5–7.5 µm. Latex white and unchanging.

HABITAT AND ROLE: This fungus is almost always found on sites that are at slightly higher elevations than the surrounding landscape and often occurs in association with dwarf birch on rocky outcrops or prominences called granite tors or hillsides having fairly well-drained soils.

EDIBILITY, TASTE, AND ODOR: Described by several authors as poisonous (gastrointestinal upsetting), this mushroom has a very biting (acrid) taste, but a fairly mild odor.

FAMILY: Russulaceae

GENUS AND SPECIES: *Russula aeruginea* Fr. (1863) (Figs. 65 and 66)

COMMON NAME: Tacky green russula

STRIKING FIELD CHARACTERS: Medium-sized, grayish green cap, white stalk, peeling cap cuticle, brittle to crumbly cap and stipe tissues, subdistant (more or less separated), even, white to cream-colored gills.

Figure 65. *Russula aeruginea* Fr. (1863)

Figure 66. *Russula aeruginea* Fr. (1863)

MACRO AND MICRO DESCRIPTION: Cap 3.5–10 cm broad, convex to convex-depressed in age; smooth, moist, tacky to viscid (sticky) in wet weather; grayish green, darker and brighter over recessed disc, becoming lighter, washed out, light olive to greenish white over expanding cap; undulating to wavy and often with split margin in age; cap tissue white, fragile to brittle and crumbly, thin; cuticle peels from cap. **Gills** attached, notched, close to spaced, broad, white becoming cream white, and even. **Stalk** 3.5–6.5 cm long, 0.7–2 cm broad; smooth, dull white to greenish white, uneven to even, subclavate with fluted apex and slightly expanded base. **Spores** creamy to dull deep yellow in deposit, subglobose, with amyloid warts and ridges, 6–9 x 5–7 µm.

HABITAT AND ROLE: On ground and most often projecting through thick moss carpets in mixed birch, white spruce and aspen forests.

EDIBILITY, TASTE, AND ODOR: This fairly easily recognized fungus is a good edible as reported by several authors.

FAMILY: Russulaceae

GENUS AND SPECIES: *Russula* aff. *emetica* (Schaeff.) Pers. (1796) (Fig. 67)

COMMON NAME: Emetic russula, the sickener

STRIKING FIELD CHARACTERS: Medium to small, deep forest floor associate of mixed white spruce and aspen forests; smooth tacky; purple-red, red to reddish white cap, white gills and white stalk, strong radish flavor.

MACRO AND MICRO DESCRIPTION: Cap 2–6 cm broad, convex, plane to irregularly depressed; deep purplish red, red to bright red, fading in age to yellowish buff (ochraceous) red to almost reddish-tinted buff over glabrous and smooth peeling cuticle and pellucid-striate margin; context white, thin, brittle to crumbly. **Gills** narrow (2–4 mm), close, even, white to yellowish white and attached, but notched. **Stalk** 2.5–5cm long, 6–14 mm broad, subclavate, smooth, dry to grayish white to same color as the gills. **Spores** white in deposit, elliptical, amyloid warts and ridges, 7–11 x 6.5–9 µm.

Figure 67. *Russula* aff. *emetica* (Schaeff.) Pers. (1796)

HABITAT AND ROLE: Terrestrial, single to gregarious, this is another suspected mycorrhizal fungus. This common "species" is found on moist to wet rich humus soil of mixed conifer-hardwood forests throughout the Alaska interior during summer (mid- to late July) and is usually associated with understory shrubs, mostly lingonberry.

EDIBILITY, TASTE, AND ODOR: This is a mild-smelling, very acrid-tasting mushroom; *Russula emetica* should not be eaten. This fungus is closely related to many other red-to-pinkish-red-capped and white-stalked *Russula* with an acrid flavor, none of which should be prepared for the table as they may induce mild to severe vomiting (as in an emetic). Our fungus is closely related to *R. montana* Shaffer and *R. mairei* Sing. Even though our species in the Interior is not directly associated with *Sphagnum* moss (as is often the case for *R. emetica*), we still think it is closer to *R. emetica* than to either of the other species.

FAMILY: Russulaceae

GENUS AND SPECIES: *Russula nana* Killerm. (1936) (Fig. 68)

COMMON NAME: Purple-red alpine russula

STRIKING FIELD CHARACTERS: Small to very small, purple-red to dull blood red with brownish red tints in age; white gills and stalk; associated with alpine dwarf willows and species of bearberry (*Arctostaphylos*).

MACRO AND MICRO DESCRIPTION: Cap 0.6–3.5 cm broad, convex, broadly convex to uneven plane with slightly recessed disc; deep purple-red to dull blood red, reddish brown in age over margin, smooth, glabrous, dry to moist, but not tacky; flesh white, pinkish red under cuticle, brittle. **Gills** white to creamy white, even, narrow, close to spaced. **Stalk** 1–2.5 cm long, 5–12 mm broad; flesh white, solid to stuffed; expanding toward base, smooth, moist, brittle. **Spores** white in deposit, broadly elliptical, amyloid warts and ridges, 7.5–9 x 6.5–7.5 μm.

Figure 68. *Russula nana* Killerm. (1936)

HABITAT AND ROLE: This small arctic to subarctic alpine *Russula* is close to *R. emetica* and *R. mairei* in appearance, but not in terms of its ecological associations. *R. alpina* is typically found in close association with *Sphagnum* moss and *R. mairei* is found on humic soils, but both are found within conifer forests in Interior Alaska. *R. nana* (= *R. alpina*) is generally relegated to high subarctic and arctic alpine meadows where dwarf willows or the heath (*Arctostaphylos alpina*) are abundant.

EDIBILITY, TASTE, AND ODOR: The combination of having the smell and taste of radishes, red color to signal "stop," and a very small size strongly suggests that this fungus is nonedible or at least nonpalatable. Neither this nor any other acrid-tasting *Russula* should be prepared for the table. Should you partake, there could be a ghastly scene about the campfire.

FAMILY: Russulaceae

GENUS AND SPECIES: *Russula* aff. *subfoetens* Wm.G. Sm. (1873) (Fig. 69)

COMMON NAME: Slightly ill-smelling russula

STRIKING FIELD CHARACTERS: Golden to yellow-brown (dull butterscotch), shiny cap, cream gill, white to faint grayish white stalk and subtly unpleasant odor.

MACRO AND MICRO DESCRIPTION: **Cap** 2.5–7.5 cm broad, convex in youth, expanding to convex-depressed to deeply depressed; viscid (tacky) to glutinous (slimy), otherwise smooth; khaki-tan in youth to orange-tan to a light burnt orange (ochraceous) color, almost that of butterscotch; margin lighter, becoming pallid (pale) to pellucid (clear, translucent, or transparent) striate (striped); flesh white, thin, appearing watery. **Gills** thick, fleshy, equal, crowded near stalk, subdistant at margin, white in youth, ivory to cream in age, broad, and brittle. **Stalk** 2.5–5 cm long, 7–17 mm broad, equal, subclavate to tapering slightly toward apex; flesh white to dull grayish white upon bruising and/or exposure to air. **Spores** are pale yellow to orange-yellow in deposit, broadly elliptical to subglobose, amyloid warts and ridges, 6–9 x 6–8 μm.

Figure 69. *Russula* aff. *subfoetens* Wm.G. Sm. (1873)

HABITAT AND ROLE: Terrestrial and a suspected mycorrhizal fungus, *Russula subfoetens* is most often found in mixed white spruce, paper birch, or quaking aspen forests in close association with blueberry, *Vaccinium uliginosum*. It is single to gregarious, plentiful some years, but not always so. It is easily spotted with its "brightly" colored cap against the dark moist humic soil of dense forests. The identification of this fungus is questionable. It clearly does not belong to either the *R. decolorans* or *R. alutacea* groups, as there is no "red" in the cap color at any time, even though a washed-out *R. decolorans* does look a lot like our fungus. It clearly is not in the *R. claroflava* group, as the brighter yellows simply are not there either. Even though our fungus has a golden brown to butterscotch cap color in age, this is no reason to suggest the potential for brighter yellow in youth. It simply is not there, as young caps are buff to khaki-tan with only hints of yellow present. We have placed it in the *R. fragrantissima* group. It has all the right colors and stature, but the Alaska interior material just does not smell all that bad, hence the reason we gave it the *R. subfoetens* label. But it is also very close to *R. laurocerasi* as well, but without the smell. It could also be in the *R. sororia* (i.e., *R. pectinatoides*) group, but the dark brown and grayish yellow tones do not appear to us. Stalk tissue in our material turns a dirty white to grayish when cut and the gill tissue is not peppery tasting.

EDIBILITY, TASTE, AND ODOR: Due to the tremendous amount of work that needs to be done on these northern *Russula* species, we recommend that none be eaten. We know very little about the toxins, and even the taxonomy is in question.

FAMILY: Tricholomataceae

GENUS AND SPECIES: *Armillaria ostoyae* (Romagn.) Herink (1973) (Fig. 70)

COMMON NAME: Honey mushroom, honey-colored mushroom

STRIKING FIELD CHARACTERS: Cap "honeydew-drop" spotted, tacky, fibrillose, with a depressed disc, veil on the stalk, in clumps or clusters (caespitose) most often at the base of stumps or living trees.

MACRO AND MICRO DESCRIPTION: **Cap** 2.5–9 cm broad, convex to broadly convex, umbonate and finally with depressed disc; tacky-sticky to slippery in wet weather; yellow tan with orange-tan clusters of erect to prostrate fibrils; margin substriate and hygrophanous (wetlike below, dry above); flesh moderate to thin, whitish to buff with brown discolorations or stains in age. **Gills** short decurrent (running down onto stalk), spaced, thin, broad, with tiers of lamellulae (small gills that do not reach from cap edge to stalk), dingy buff, to light yellow-tan with darker brown spots in age. **Stalk** 5–8 cm long, 10–25 mm thick, tapering up to veil and equal to slightly flaring at apex above veil; dry whitish buff to tan streaked. Veil soft, defined, cottonlike, dull white and persistent. **Spores** white in deposit, elliptical, smooth, non-amyloid, 7–11 x 5–6 µm.

Figure 70. *Armillaria ostoyae* (Romagn.) Herink (1973)

HABITAT AND ROLE: This fungus is known to be a virulent root rotter as well as simply a decomposer of woody substrates. It often sends out thick brown rootlike rhizomorphs that link it to other suitable substrates. *Armillaria ostoyae* almost always occurs in caespitose (aggregated but not fused) clusters of a few to many fruiting bodies joined to others at a common stalk base. We generally find it at the base of aspen or balsam poplar, rarely on birch, but sometimes on old white spruce stumps.

EDIBILITY, TASTE, AND ODOR: *Armillaria ostoyae* is considered edible and choice to edible with caution. Taste and odor are mild to slightly fragrant (sweet). This fungus represents a conglomerate of species in North America and Europe. The jury is still out on species separation and so we are unsure where taxonomically the Alaska interior form will finally reside. If you eat the fungus, do so with caution as this complex has been known to cause gastrointestinal upset in some.

FAMILY: Tricholomataceae

GENUS AND SPECIES: *Cantharellula umbonata* (J.F. Gmel.) Singer (1936) (Figs. 71 and 72)

COMMON NAME: Northern moss clitocybe

STRIKING FIELD CHARACTERS: Cap light bluish gray, bruising darker, umbonate; gills whitish cream, crowded, narrow, forked, stain reddish brown in age or when damaged; stalk base with silky white mycelium, in and around thick green moss carpets in western DNP&P.

Figure 71. *Cantharellula umbonata* (J.F. Gmel.) Singer (1936)

MACRO AND MICRO DESCRIPTION: Cap 25–32 mm, broadly convex, sharply umbonate to depressed and rounded umbonate; color gray to elephant (dark brownish gray) skin to metal gray in age or where bruised/handled; margin lighter, almost a whitish gray and inrolled; flesh water soaked-pliant, turning a watery yellow upon exposure to air; surface smooth, moist to minutely fibrillose. **Gills** cream-white, decurrent, branching into 4 tiers of dichotomously branched smooth, entire, blunt ridged, thin, close lamellae, up to 1 mm broad; color (dried) white to light cream with darkening (buff) edges. **Stalk** tapering, 4–7 mm at apex, 3–6 mm below and 2–5 mm at broadening base of silky white mycelium entwined with the moss substrate, 40–50 mm long; color silky white with light gray flushes, darkening where handled, solid, moist flexuous, pliant, smooth to fibrillose; flesh solid, somewhat

Figure 72. *Cantharellula umbonata* (J.F. Gmel.) Singer (1936)

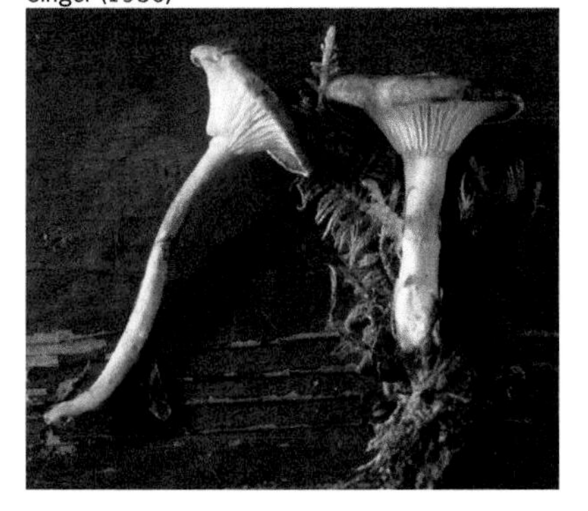

water soaked, white, no veil remains. **Spores** white in deposit, spindle-shaped, smooth, amyloid, 7–11 x 3–4 µm.

HABITAT AND ROLE: Solitary, occurring on *Hypnum* and *Dicranum* mosses, appearing in late July after prolonged rains, in open rolling tundra above margin of Wonder Lake.

EDIBILITY, TASTE, AND ODOR: This fungus is not recommended as it is so close to a number of nonedible *Clitocybe* species best left for the squirrels. However, its taste is mild and its odor fishy. The solitary nature of its growth pattern would cause one to search for hours, if not days, to find enough for a meal if you were to want to eat it.

FAMILY: Tricholomataceae

GENUS AND SPECIES: *Collybia dryophila* (Bull.) P. Kumm. (1871) (Fig. 73)

COMMON NAME: Nut-brown collybia

STRIKING FIELD CHARACTERS: Cap expanded, wavy, nut brown, white to cream attached gills and smooth yellowish tan stalk; on needle duff (litter or detritus), dead leaves, and other plant debris.

MACRO AND MICRO DESCRIPTION: **Cap** 2.5–5 cm broad, convex becoming broadly convex and finally plane (flat), but wavy; smooth, glabrous; moist but not tacky; yellow brown to a chestnut brown; margin inrolled at first becoming recurved in age and with a very thin sterile wavy margin; flesh white to cream, thin, pliant. **Gills** attached, notched, crowded, thin, narrow, white to cream. **Stalk** 2–6 cm long, 3–7 mm broad; smooth, moist; yellow-brown to tan-brown below and becoming yellowish buff to light cream at apex; hollow base often with conspicuous white mycelium. **Spores** white to pale cream in deposit, elliptical, smooth, 5–7 x 2–3.5 µm.

Figure 73. *Collybia dryophila* (Bull.) P. Kumm. (1871)

HABITAT AND ROLE: This fungus is a saprobic decomposer of forest floor litter and can be found singly to scattered to amassed in small clumps (but not joined together) on mossy forest floors under mixed aspen, birch, willow, and white spruce.

EDIBILITY, TASTE, AND ODOR: Edibility of this fungus in North America is questionable. In one instance, a whole class of students developed gastrointestinal upset from eating it. Some field guides claim it to be edible. Frankly, we would hold off on this one. Ours may not be identical to the edible European form.

FAMILY: Tricholomataceae

GENUS AND SPECIES: *Cystoderma amianthinum* (Scop.) Fayod (1889) (Fig. 74)

COMMON NAME: Unspotted cystoderma

STRIKING FIELD CHARACTERS: Cap dry, granular, cream yellow to yellow-orange; stalk flocculent (clothed with small with woolly tufts), ringed, but soon breaking up; margin with irregular veil remains; on mosses under white spruce.

MACRO AND MICRO DESCRIPTION: Cap 2–5 cm broad, convex to broadly convex becoming plane (flat) in age; dry; red-brown fibrillose scales forming broken concentric rings over a lighter yellow-orange to orange-tan background; flesh white, thin, tough-pliant. **Gills** white to cream in age, thin, broad, crowded, attached to stalk. **Stalk** 3–5 (7) cm long, 3–5 mm broad, equal, dry, with small tufted scales below ring, smooth and lighter above to apex, pliant-tough; same color as cap; veil delicate, membranous white to cream. **Spores** white in deposit, elliptical, smooth, amyloid, 4–7 x 3–4 µm.

Figure 74. *Cystoderma amianthinum* (Scop.) Fayod (1889)

HABITAT AND ROLE: This brightly colored mushroom, which is often conspicuous against the deep green of a moss-carpeted white spruce forest floor, occurs directly on the moss and is a suspected decomposer of moss stems and leaves. Single to scattered, common, in deeper, often mixed, forests and campsite settings like those found at Tetlanika.

EDIBILITY, TASTE, AND ODOR: *Cystoderma amianthinum* is reputed to be nonpoisonous, but since this is a group not well studied from Alaska, we would be inclined not to eat it. Its taste is mild, however, and odor like that of fresh ear corn. Three to four easily confused species are found in southeast Alaska and could be confused with this one.

FAMILY: Tricholomataceae

GENUS AND SPECIES: *Cystoderma fallax* A.H. Sm. & Singer (1945) (Fig. 75)

COMMON NAME: Conifer cystoderma

STRIKING FIELD CHARACTERS: Robust, dull russet (reddish brown), granular cap with marginal partial veil remains, white gills, granular stalk with annulus and smooth above to apex.

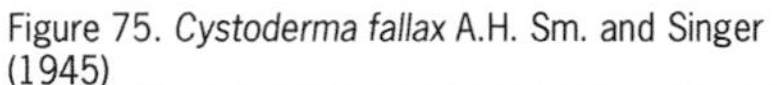

Figure 75. *Cystoderma fallax* A.H. Sm. and Singer (1945)

MACRO AND MICRO DESCRIPTION: Cap 1–3 (4) cm broad, convex, becoming broadly so and finally plane in age; granular; dull russet (reddish) brown to dull orange brown; flesh thin, white. **Gills** attached, notched, close, tiered, narrow and white. **Stalk** 2–8 cm long, 3–5 mm wide, equal to tapering slightly toward its persistent annulus and apex, granulose below, smooth above veil to apex; lighter tawny orange below to same color as cap and white above ring. Annulus smooth, white above, granular, colored as stipe below. **Spores** white in deposit, broadly elliptical to subglobose, smooth, amyloid, 3.5–5.5 x 3–4 µm.

HABITAT AND ROLE: Found on deep mats of ground-cover feather mosses (*Pleurozium schreberi*) in white spruce and mixed spruce-hardwood forests at the east end of DNP&P. The role of this fungus is that of a saprobic decomposer of moss parts.

EDIBILITY, TASTE, AND ODOR: No reports of the edibility of this fungus are known. Like *Cystoderma amianthinum*, it is perhaps best left for the squirrels. Taste is mealy (flourlike), but its odor is that of mild green fresh corn.

FAMILY: Tricholomataceae

GENUS AND SPECIES: *Laccaria laccata* (Scop.) Fr. (1884) (Fig. 76)

COMMON NAME: Common laccaria, waxy laccaria, or the old man of the woods

STRIKING FIELD CHARACTERS: Pinkish brown cap, stipe and rosy flesh pink gills; "long" flexuous and longitudinally fibrillose stalk; cap and stalk smooth to scurfy (unevenly roughened) fibrillose.

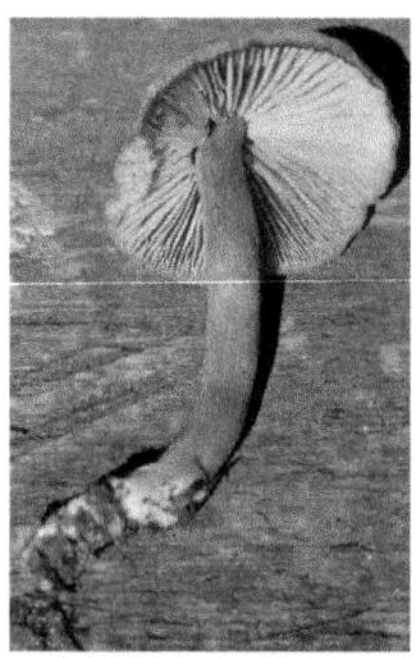

Figure 76. *Laccaria laccata* (Scop.) Fr. (1884)

MACRO AND MICRO DESCRIPTION: Cap 1–6.5 cm broad, convex to broadly convex, but never flat; moist, smooth to minutely scurfy (roughened) or rimose (cracked); margin faintly striate, wavy with a thin sterile margin (gills not reaching to edge), earthy to pinkish brown, sometimes with faint violaceous tints; flesh thin, same color as stalk. **Gills** broadly attached (adnate) to subdecurrent, rose to flesh pink, spaced, fleshly, narrow and tiered. **Stalk** 3.5–10 cm long, 6–12 mm broad, moist, smooth to scaly, unequal and flexuous, longitudinally striate, same color as cap to lighter. **Spores** white in deposit, globose, spiny, 7–10 x 6–9 µm.

HABITAT AND ROLE: This is a known ectomycorrhizal fungus found chiefly in white spruce forests on moss-covered soil and other types of conifer forests to the south.

EDIBILITY, TASTE, AND ODOR: This is reported as being a good edible, but with caution. No known toxins have been reported, but then, that is not necessarily saying very much.

FAMILY: Tricholomataceae

GENUS AND SPECIES: *Melanoleuca melaleuca* (Pers.) Murrill (1911) (Fig. 77)

COMMON NAME: Changeable tricholoma

STRIKING FIELD CHARACTERS: Large, flat to depressed, brown to chestnut brown and smooth cap, cream to buff, and crowded gills on a long, hollow stalk, occurring in alpine meadows.

MACRO AND MICRO DESCRIPTION: Cap 5–13 cm broad, broadly convex, expanding to plane, disc a low umbo becoming depressed in age; dark tan to brown, mostly deeper brown, chestnut to chocolate brown; moist, smooth, and appearing almost zonate (ringed color highlights); extreme marginal edge tan to lighter cream, straight, wavy in age; flesh thick, white to buff, pliant. **Gills** white, cream, buff to very light tan in age, very crowded, thin, broad, attached (notched) to broadly so. **Stalk** 5–13 cm long, 8–15 mm broad, white or buff to light cream, smooth; flesh stuffed to hollow, otherwise stalk brittle, fibrous, equal to slightly swollen toward base. **Spores** white in deposit, elliptical, minutely warted, amyloid, 6–8 x 4–5.5 µm.

Figure 77. *Melanoleuca melaleuca* (Pers.) Murrill (1911)

HABITAT AND ROLE: This fungus is one of the largest in size to be seen in higher alpine meadows, where it occurs in association with mosses, ericaceous shrubs, and dwarf willows. It very well could be a mycorrhiza former, but is thought also to be a saprobic decomposer of the plant debris to which it always seems to be attached.

EDIBILITY, TASTE, AND ODOR: This fungus is described as being edible or with unknown edibility. When collected, *Melanoleuca melaleuca* often becomes so watery and heavy to fragile that it is hard to get it home for the table in any kind of shape to be considered edible. Taste and odor are both mild.

FAMILY: Tricholomataceae

GENUS AND SPECIES: *Microcollybia tuberosa* (Bull.) Lennox (1979) (Fig. 78)

COMMON NAME: Tuberous collybia

STRIKING FIELD CHARACTERS: Very small, white cap, long thin equal (terete [rounded cylindrical]) stalk, from a reddish brown wrinkled tuber (sclerotium) at the base which is embedded in substrate, such as an old rotting mushroom.

MACRO AND MICRO DESCRIPTION: **Cap** 3–15 mm broad, convex; white to buff over disc; moist, smooth, with a sheen; margin inrolled at first; flesh thin, white. **Gills** thin, close, narrow, attached, white to buff. **Stalk** 2.5–5 cm long, 0.5–2 mm broad, white to ivory, pliant, flexuous, narrowing slightly toward base where it arises from a long (2–8 mm), wrinkled, tan, light brown to dark chocolate brown tuber. **Spores** white in deposit, elliptical, smooth, 3–6 x 2–3 µm.

Figure 78. *Microcollybia tuberosa* (Bull.) Lennox (1979)

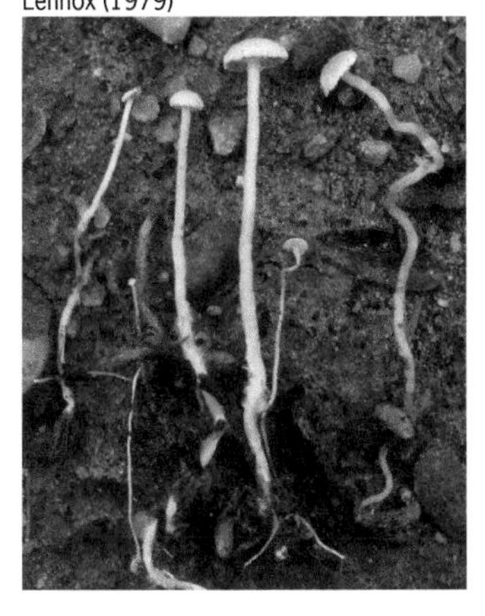

HABITAT AND ROLE: *Microcollybia tuberosa* is very conspicuous against a wet, dark forest floor in late summer to early fall, where it occurs on the often-decomposing blackish ooze of another old mushroom fruiting body, which, more than likely, will be one representing the families Boletaceae or Russulaceae. This fungus is a saprobic decomposer. It possesses an often inconspicuous but expanded seedlike tuberous growth at the stalk base that is most frequently buried in the substrate on which it is growing.

EDIBILITY, TASTE, AND ODOR: This species is reported to be nonpoisonous, but its size would necessitate a lot of collecting before you would even have one mouthful.

FAMILY: Tricholomataceae

GENUS AND SPECIES: *Mycena pura* (Pers.) P. Kumm. (1871) (Fig. 79)

COMMON NAME: Pink or clean mycena

STRIKING FIELD CHARACTERS: A "large" *Mycena* with fleshy pink, rose, or purplish hues; round to flattened (even subfluted) stalk; and "showy" violaceous, spaced, and thick gills; associated with mixed spruce-hardwood forests.

Figure 79. *Mycena pura* (Pers.) P. Kumm. (1871)

MACRO AND MICRO DESCRIPTION: Cap 1–4.5 cm broad, broadly convex, low umbo over disc, to plane and finally with a recurved (upturned) and wavy translucent-striate margin; smooth, moist to somewhat translucent when wet; dull flesh to rose and finally lilaceous tinted to purple. **Gills** attached (notched), close in youth to expanded, thick, fleshy, and spaced in age, broad; pinkish flesh to rosy lilaceous (clean-looking). **Stalk** 3.5–10 cm long, 3–8 mm broad, round in youth to compressed (almost fluted) in age, equal to becoming enlarged toward whitish base, often twisted, otherwise smooth and moist; color similar to that of the cap. **Spores** white in deposit, broadly elliptical to cylindrical, smooth, amyloid, 5–9 x 3–4 µm.

HABITAT AND ROLE: This fungus is a saprobic decomposer of organic matter (needles, mosses, etc.) that occurs as part of the organic ground cover and is found in mixed white spruce and aspen forests. *Mycena pura* is perhaps the most variable fungus in terms of color, stalk length, and often contorted cap shapes that we have witnessed in interior Alaska. It is fairly common, however, and found singly to gregarious and mostly associated with wet deep moss carpets on the forest floor.

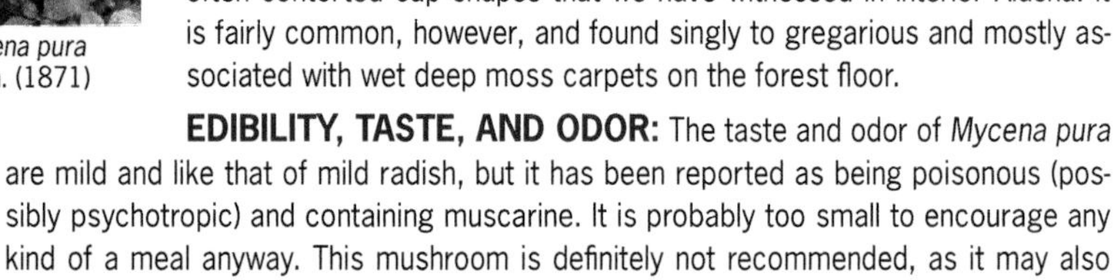

EDIBILITY, TASTE, AND ODOR: The taste and odor of *Mycena pura* are mild and like that of mild radish, but it has been reported as being poisonous (possibly psychotropic) and containing muscarine. It is probably too small to encourage any kind of a meal anyway. This mushroom is definitely not recommended, as it may also cause gastrointestinal upset in addition to other symptomatic conditions of muscarine poisoning.

FAMILY: Tricholomataceae

GENUS AND SPECIES: *Phyllotopsis nidulans* (Pers.) Singer (1936) (Fig. 80)

COMMON NAME: Orange mock oyster, nest-cap mushroom

STRIKING FIELD CHARACTERS: This is a relatively small orange, dry velvety "shelf or bracket" fungus with yellow-orange gills and no stalk.

MACRO AND MICRO DESCRIPTION: Cap 2.5–8cm broad, semicircular, broadly convex to plane, densely hairy upper surface; buff to orange-yellow to yellowish orange; margin inrolled, scalloped; flesh paler than cap; dry. **Gills** attached to hairy white base, close to crowded, narrow to moderately broad; yellowish buff, orange-yellow to burnt orange in age. **Stalk** absent. **Spores** pale pinkish to pinkish brown in deposit, sausage-shaped, smooth, 5–8 x 2–4 µm.

Figure 80. *Phyllotopsis nidulans* (Pers.) Singer (1936)

HABITAT AND ROLE: Single to overlapping clusters are found on alder and white spruce, which it decomposes.

EDIBILITY, TASTE, AND ODOR: Not known to be poisonous or its edibility reportedly unknown, *Phyllotopsis nidulans* is perhaps best left alone as its odor is disgusting, strong, or disagreeable, like hydrogen sulfide (H_2S) gas (rotten eggs) to mild. The taste is unknown to us.

FAMILY: Pluteaceae (= Volvariaceae)

GENUS AND SPECIES: *Pluteus cervinus* P. Kumm. (1871) (Fig. 81)

COMMON NAME: Fawn-colored pluteus, deer mushroom

STRIKING FIELD CHARACTERS: Whitish to pinkish tan to light grayish-tan cap broadly convex, the flesh pinkish, with free gills, and typically fruiting from downed birch logs.

Figure 81. *Pluteus cervinus* P. Kumm. (1871)

MACRO AND MICRO DESCRIPTION: Cap 3.5–11.5 cm, broadly convex; smooth, moist to somewhat sticky; often with appressed fibrils over disc; margin becoming recurved and split; context (flesh) moderately thick, ivory, becoming dirty white. **Gills** free, fairly close, broad, brittle and easily damaged; whitish tan becoming flesh pink at maturity. **Stalk** 3.5–7.5 cm long, 5–10 mm broad, longitudinally fibrous and slightly "candy cane" twisted; firm, pliant, solid, smooth, and whitish tan or ivory. **Spores** flesh to pinkish brown in deposit, elliptical, smooth, 5–8 x 4–6 µm.

HABITAT AND ROLE: In the interior of Alaska, this fungus is fairly common. It seemingly always occurs on hardwoods, including birch logs well on their way to being decomposed and delignified. *Pluteus cervinus* is a saprobe and, we suspect, a cellulose decomposer, given the spongy condition hardwood logs are in when we see this fungus fruiting on them. Rarely do we find more than two on any one rotting log, and most often only a single fruiting body is present.

EDIBILITY, TASTE, AND ODOR: *Pluteus cervinus* is reported to be edible. We have never eaten this fungus as it has never been found in abundance, and when found is often too mature for the frying pan. Taste is mild to faintly metallic when fresh. Odor is pleasant. This fungus, given its fruiting locality, pinkish flesh-colored gills, and candy-cane twist to its stalk, can be clearly separated from other mushrooms found on similar woody substrates.

FAMILY: Paxillaceae

GENUS AND SPECIES: *Paxillus involutus* (Batsch) Fr. (1838) (Fig. 82)

COMMON NAME: Poison paxillus, inrolled or involute paxillus

STRIKING FIELD CHARACTERS: Drab orange, dry, feltlike to shiny cap with inrolled hairy margin; stalk off-center, short; gills dull yellowish orange.

MACRO AND MICRO DESCRIPTION: **Cap** 3.5–12.5 cm broad, broadly convex to flat and becoming depressed early; slimy capped in wet weather, otherwise dry matted felty (fibrillose); dull yellowish brown, often spotted; splitting (rimose) near persistently inrolled and fuzzy margin. **Gills** narrow, short decurrent, crowded, forked once, yellow-tan to yellow-orange, bruising reddish orange to brown. **Stalk** 12–25 mm long, 10–13 mm broad (short and stocky), dry, and same color as cap to slightly lighter. **Spores** yellow-brown to brown in deposit, elliptical, smooth, 7–10 x 4–6 μm.

Figure 82. *Paxillus involutus* (Batsch) Fr. (1838)

HABITAT AND ROLE: Terrestrial, solitary, scattered to gregarious on rich humus of alpine tundra peat or mixed white spruce–paper birch forests, and a suspected mycorrhizal associate, if not with paper birch, then with ericaceous understory plants such as crowberry. We have collected this from low ericaceous shrub/willow/dwarf birch–dominated tundra of high alpine sites near large granite tors north of Coldfoot on the Dalton Highway, clearly some distance from any "forests" as we think of them.

EDIBILITY, TASTE, AND ODOR: The mild flavor and fungoid odor of *Paxillus involutus* may deceive, as this fungus is generally regarded as a gastrointestinal irritant and a liver destroyer and should never be eaten. In Europe and parts of North America it has been reported as being edible and is enjoyed by some, even though *liver toxicity is well documented.* Do not partake of this mushroom as it is reported to possess carcinogenic toxins.

FAMILY: Cortinariaceae

GENUS AND SPECIES: *Cortinarius* (*Sericeocybe*) *alboviolaceus* (Pers.) Fr. (1838) (Fig. 83)

COMMON NAME: Light, pale, or silver-violet cortinarius

STRIKING FIELD CHARACTERS: Silky white cap with lilac to violet tints; dry, rusty brown gills.

MACRO AND MICRO DESCRIPTION: **Cap** 1.5–5cm broad, hemispheric, bell-shaped to rounded bell-shaped, convex to plano-convex (only slightly convex); surface dry, silky, shiny; color pale lilac-white to pale violet, becoming silvery violet, lilac-white or even white; flesh pallid (pale, whitish) to pale violet to rusty brown and "marbled";

margin inrolled and sheathed in a thick cobweblike (fibrillose) veil that is the same color as the cap. **Gills** are notched to broadly attached, close to somewhat separated, thick, broad, uneven; clay-violet to purple-gray and eventually becoming cinnamon to rusty-brown. **Stalk** 4–7 cm long, 1–2 cm broad, same color as the cap and often with a rust-stained base; long-clavate (club-shaped), smooth, dry; flesh pale violet to rusty brown and becoming marbled. Veil whitish, cobweblike and soon disappearing (evanescent). **Spores** rusty brown in deposit, elliptical, minutely roughened, 7–10 x 4–6 µm.

Figure 83. *Cortinarius* (*Sericeocybe*) *alboviolaceus* (Pers.) Fr. (1838)

HABITAT AND ROLE: Terrestrial, single, scattered to mostly gregarious or in small clumps in mixed birch–aspen–white spruce forests during late summer. This is an ecto mycorrhizal fungus.

EDIBILITY, TASTE, AND ODOR: Some authorities report edibility as unknown, while others report this species as being edible, but not usually pleasant tasting. It tastes earthy and often watery. The insects love them in Alaska.

FAMILY: Cortinariaceae

GENUS AND SPECIES: *Cortinarius* (*Dermocybe*) aff. *croceus* (Schaeff.) Gray (1821) (Fig. 84)

COMMON NAME: Olive yellow cortinarius

STRIKING FIELD CHARACTERS: Yellow to reddish orange cap; chrome yellow gills; yellowish white to citron (lemon) yellow stalk with rusty red fibrils and yellowish orange cobweblike veil.

MACRO AND MICRO DESCRIPTION: **Cap** 2.5–4 cm broad, hemispheric, parabolic, pulvinate (cushion-shaped) to convex, finally plano-convex and often with a low umbo; margin incurved at first, then expanding; yellow-brown (olivaceous) reddish orange to red-brown, paler at margin to pale yellow and same color as stalk; dry, smooth to minutely fibrillose; flesh pale yellow, darkest at stalk base, and with a lilaceous (pale pinkish purple) tint. **Gills** close but not crowded, chrome yellow, ochraceous yellow to orange-yellow, uneven, attached (notched). **Stalk** 3–4.5 cm long, 4–10 mm broad, cylindrical, hollow, equal to slightly expanded base; golden yellow to pale olivaceous yellow, with red to red-orange fibrils; basal mycelium cream-colored. Veil yellow to yellow-orange, fibrillose and cobweblike. **Spores** rusty brown in deposit, 7–8 x 4–4.5 µm.

HABITAT AND ROLE: Terrestrial, solitary, scattered to gregarious, this fungus grows in mixed aspen–birch–white spruce forests on soil of the forest floor in association with ericaceous plants such as crowberry, small-leafed Labrador tea, and blueberry. It is a mycorrhizal species and has many close "relatives" one is apt to find close by. A similar fungus with a brown cap (without the red-orange cap color) is *Cortinarius croceus* subsp. *croceus*; an olive cap/stalk tissue is found in *C. croceus* subsp. *norvegicus* (a

Figure 84. *Cortinarius* (*Dermocybe*) aff. *croceus* (Schaeff.) Gray (1821)

European species), which has brighter yellow gills and a more cinnamon cap. *C. croceofolius*, with saffron to rusty yellow or orange gills and a more cinnamon-colored cap, is another similar species, and there also is the more distant *C. cinnamomeus/C. cinnamomeo-luteus* complex with more olive tints over the cap and stalk, but burnt orange, rusty to reddish orange gills. The *C. cinnabarinus/C. sanquineus* group has blood red to reddish purple-colored gills, but may be found in similar habitats in interior Alaska.

EDIBILITY, TASTE, AND ODOR: All of these relatively small, brightly colored Cortinarii should be avoided. There are some really bad actors in closely related forms such as *C. orellanus* and *C. speciosissimus* (not previously documented in North America) and *G. gentilis*, all of which are deadly poisonous. The orellanin poisoning by these fungi may have up to a *two-week* delay before symptoms appear! To the untrained and the trained eye alike, these fungi are difficult to separate with any level of confidence using only macroscopic characters. The taste is mild to slightly metallic and the odor is pleasant.

FAMILY: Cortinariaceae

GENUS AND SPECIES: *Cortinarius* aff. *mucosus* (Bull.) Cooke (1867) (Fig. 85)

COMMON NAME: Shiny cortinarius

STRIKING FIELD CHARACTERS: Tawny, orange to reddish brown, smooth glutinous (sticky) cap, lighter margin; long, straight, white slimy stalk.

Figure 85. *Cortinarius* aff. *mucosus* (Bull.) Cooke (1867)

MACRO AND MICRO DESCRIPTION: Cap 2–6 cm broad, convex to plane or expanded, umbonate, uplifted or depressed in age; smooth, glutinous (sticky); orange brown, chestnut brown to reddish brown, becoming paler in age and over margin; flesh whitish cream to buff. **Gills** close, notched, off-white in youth, becoming pale ochraceous to orangish tan, finally browning to rusty cinnamon. **Stalk** 8–12 cm long, 1–2 cm broad, equal (cylindrical), tapering to apex, irregular at base, shiny, white faint lavender tint and bruising buff, slimy, but not scaly belted. Veil is whitish tan, glutinous (sticky), disappearing early with cinnamon spore-colored remains and collapsed fibrous cobweblike cortina about stalk near the apex. **Spores** rusty brown in deposit, long elliptical, roughened, 11–18 x 5–7.5 µm.

HABITAT AND ROLE: *Cortinarius* aff. *mucosus* is terrestrial (mycorrhizal), scattered to gregarious in coniferous spruce forests at or near treeline, where it is associated with reindeer lichens (*Cladonia rangiferina*) and mountain avens *(Dryas octapetala)*. This fungus has also been seen far to the north (Brooks Mountain Range) in Alaska associated with dwarf birch and willows as well as with *Dryas* along the Dalton Highway. This fungus will bring other look-alikes to mind as well. At first sight, *C. mucosus* is strikingly similar to *C. collinitus* and *C. trivialis*, but lacks the brown scaly belts or fibrous band on the stalk of the latter and nearly colorless to faint lilac glutinous bands of the former. *C. muscigenus* is reported as being similar, but this species also lacks glutinous bands on the stalk. *C. septentrionalis*, a northern (European) deciduous forest species, is usually darker and more reddish brown. Once again, one must go to microscopic (spore) characters to obtain a positive identification.

EDIBILITY, TASTE, AND ODOR: This is one more "cort" best left alone. Its taste is mild, not bitter, and odor is fungoid (decidedly mushroomlike).

FAMILY: Cortinariaceae

GENUS AND SPECIES: *Cortinarius trivialis* J.E. Lange (1940) (Fig. 86)

COMMON NAME: Common or belted slimy cortinarius

STRIKING FIELD CHARACTERS: Yellow, dull to reddish tan/brown slimy cap; ivory, buff to pale whitish attached gills; slimy white stalk above, breaking up into yellow-brown bands below to base; cobweblike veil disappearing early.

MACRO AND MICRO DESCRIPTION: **Cap** 2–7.5 cm broad, campanulate-convex (convex bell-shaped), convex, flattened-convex, broadly convex, umbonate, plane to slightly depressed; smooth, slimy to viscid; yellowish orange, yellow-brown, orange-brown, tawny, ochraceous, or even with reddish brown tints; margin inrolled at first, lighter, expanding in age, but not striate in our northern species; flesh firm, ivory to pale buff. **Gills** close, notched, whitish brown to brown, becoming rusty brown in age. **Stalk** 3.5–9cm long, 5–15 mm broad, slimy to sticky, subequal to tapering toward base; white above, breaking up into irregular yellowish tan, ochraceous to rusty brown bands or scaly rings below to base, lacks violaceous tints but hints of lavender are somewhat evident. Veil fibrillose, cortina whitish, forming a "ring" of hairs collapsed against stalk and often colored with the rusty brown spores. **Spores** rusty brown in deposit, variable, 10–15 x 7–8 µm.

Figure 86. *Cortinarius trivialis* J.E. Lange (1940)

HABITAT AND ROLE: Terrestrial and mycorrhizal, this mushroom is scattered, gregarious to clustered during peak fruiting periods in mixed birch-aspen-conifer woodlands in the eastern portion of DNP&P. Several look-alikes are distinguished primarily on the presence or absence of lavender to violaceous tints in the cap, gills, stalk sheath, and flesh (apex primarily). Spore size, shape, and ornamentation, which one cannot determine in the field, are also important diagnostic features. Other similar species that come to mind are *Cortinarius collinitis* and/or *C. mucifluus*. *C. elatior* lacks striking stalk "bands" and may be more closely associated with aspen.

EDIBILITY, TASTE, AND ODOR: The edibility of this fungus, which is listed as *Cortinarius collinitus* subsp. *trivialis* by some authors, is unknown. Some claim it to be edible or that it can be eaten, whatever that might really mean. The taste is nutty and its odor is mildly bitter to fungoid; in other words, not too tasty and eating is not worth the risk.

FAMILY: Cortinariaceae

GENUS AND SPECIES: *Crepidotus mollis* (Schaeff.) Staude (1857) (Fig. 87)

COMMON NAME: Soft or jelly crepidotus

STRIKING FIELD CHARACTERS: Broadly convex to flattened small cap with inrolled margins, on woody hardwood twigs, stalk absent, and with rusty tan gills.

MACRO AND MICRO DESCRIPTION: Cap 0.5–2.5 cm broad, convex to broadly convex, becoming almost plane (flat); dry, buff white to light tan, minute brown fibrous scales over disc, lighter over velvety white inrolled margins. **Gills** are somewhat separated, narrow, buff to cinnamon tan. **Stalk** absent, attachment to the substrate, eccentric (off-centered). Flesh is buff, thin. **Spores** yellow-brown to dull brown in deposit, elliptical, smooth, 7–11 x 4.5–6.5 µm.

Figure 87. *Crepidotus mollis* (Schaeff.) Staude (1857)

HABITAT AND ROLE: We typically see this small and often inconspicuous fungus in small overlapping clusters on small sticks of alder, willow, or cottonwood on or near the ground in a jumble of moist forest floor woody debris. Its role is that of saprophytic decomposer.

EDIBILITY, TASTE, AND ODOR: The edibility of *Crepidotus mollis* is reported as being unknown; its taste is mild metallic (like that of iron) and its odor is fresh and woody. Do not confuse this with the small oyster mushroom (*Pleurotus ostreatus*) that is typically eaten. The latter is usually white to cream at maturity and most often much larger, and having white spores.

FAMILY: Cortinariaceae

GENUS AND SPECIES: *Galerina* aff. *paludosa* (Fr.) Kühner (1935) (Fig. 88)

COMMON NAME: Sphagnum galerina

STRIKING FIELD CHARACTERS: Buff, tan, brown to reddish brown; pale striate margin; fibrillose; thin, fragile stalk; on wet *Sphagnum* moss mounds.

MACRO AND MICRO DESCRIPTION: Cap 10–30 mm broad, conic, almost cuspidate (ending in a sharp point), but most often umbonate (with a central cone-shaped projection); buff, ochre, yellow-tan to reddish brown, pale striate, to primrose yellow and smooth in age, moist. **Gills** relatively thick, spaced, uneven, narrow, adnexed (narrowly joined to the stalk), adnate (broadly attached) to very short decurrent, light yellow to rusty brown. **Stalk** 4–7 cm long, 2–3 mm broad, equal, hollow, grayish tan, ochre to yellow-brown and often lighter than cap color; buff to whitish tan fibrils over most of

Figure 88. *Galerina* aff. *paludosa* (Fr.) Kühner (1935)

stalk, and fragile. **Spores** dark rusty brown, rugulose to smooth, 10–13 x 6–7 µm.

HABITAT AND ROLE: This sphagnicolous (*Sphagnum*-loving) *Galerina* is one of a group of fungi very difficult to distinguish from one another in the field and often identifiable only with a microscope, but which abound at these northern latitudes in the upper boreal forest and on gently sloping and undulating hillsides where large deep hummocks of *Sphagnum* moss are found. This fungus belongs to or is close to other Galerinas represented by *G. allospora, G. calyptrata, G. fallax, G. hypnorum, G. lubrica, G. mniophyla, G. sphagnicola, G. sphagnorum,* and others. These LBMs (little brown mushrooms) must be examined microscopically for a number of characters inherent in their spores, stalk and cap cell makeup, and hyphal features. They are all decomposers of their dead moss substrates, in this case *Sphagnum* moss.

EDIBILITY, TASTE, AND ODOR: Species of *Galerina* should be totally avoided and never eaten, especially this one. There are some bad actors in this hard-to-identify group of LBMs. Larger forms than the ones pictured here, such as *G. autumnalis, G. marginata,* and *G. venenata,* are reported to possess deadly toxins. The *Galerina* pictured had a mild taste and fishy odor, probably due to its substrate. We would never have eaten this fungus! We merely tasted it for scientific and descriptive purposes.

FAMILY: Cortinariaceae

GENUS AND SPECIES: *Inocybe geophylla* var. *geophylla* (Pers.) P. Kumm. (1871) (Fig. 89)

Figure 89. *Inocybe geophylla* var. *geophylla* (Pers.) P. Kumm. (1871)

COMMON NAME: Little white inocybe, white fiber head

STRIKING FIELD CHARACTERS: Small, totally white, dry, fibrillose, with a conical to knobbed cap and slightly swollen stalk base, occurring in deep mosses.

MACRO AND MICRO DESCRIPTION: Cap 0.5–1 cm broad, conical to campanulate (bell-shaped), umbonate to subconvex, silky fibrillose, white, margin inrolled. **Gills** dull white, dirty gray to tan, notched, crowded, uneven, thin, and narrow. **Stalk** is 3–5 cm long, 2–4 mm broad, cylindrical, flexuous, more or less equal, slightly swollen at base, pliant, satinlike (shiny). Veil cobweblike and soon disappearing. Flesh white. **Spores** brown, elliptical, smooth, 8–10 x 5–6 µm.

HABITAT AND ROLE: *Inocybe geophylla* is solitary, scattered to occasionally gregarious, on soil, usually associated with a moist moss cover in spruce forests; a mycorrhiza former.

EDIBILITY, TASTE, AND ODOR: Though very small, this is a poisonous fungus. *Inocybe geophylla* should never be eaten, as it contains muscarine. Its taste is somewhat metallic and the odor is spermatic.

FAMILY: Strophariaceae (Cortinariaceae)

GENUS AND SPECIES: *Pholiota aurivella* (Batsch) Fr. (1871) (Fig. 90)

COMMON NAME: Golden or slimy pholiota

STRIKING FIELD CHARACTERS: Slimy in wet weather, to viscid (sticky or slippery), tacky or even "smooth" in dry or drier weather; yellow-orange to tawny; covered with large "glued down, spotlike" scales; stalk yellow to yellow-brown, annulate, dry-scaly below and becoming dense toward its base.

MACRO AND MICRO DESCRIPTION: **Cap** 2–5 cm broad, broadly campanulate (bell-shaped) to convex with a broad disc; shiny in wet weather, otherwise viscid (sticky), tacky to dry and with a sheen; pale to dark yellow, tawny, golden orange to a light rusty orange and decorated with most often flattened, triangular, rusty brown scales that may "wash" off with age; margin often with veil remnants; flesh off-white to yellow. **Gills** broad, close, broadly attached to notched, whitish to yellow, but becoming rusty brown in age. **Stalk** 5–8 cm long, 0.8–1.5 cm broad, equal to tapering slightly toward the tuftlike base; dry, scurfy to fibrillose, generally smooth above veil and scaly below; cream, pale yellow to yellow-brown with burnt orange scales. Veil fibrillose, whitish to pale yellow or light olive-yellow, forming a slight but early disappearing stalk "ring," which sometimes may be even more obvious on the margins of young caps. **Spores** brown, with a germ pore, elliptical, smooth, 8.5–10 x 5–6 µm.

Figure 90. *Pholiota aurivella* (Batsch) Fr. (1871)

HABITAT AND ROLE: This lignicolous (wood-inhabiting) fungus grows in tufts or clusters on dead aspen in mixed hardwood and white spruce forests.

EDIBILITY, TASTE, AND ODOR: This fungus belongs to a complex species, the members of which are difficult to field identify. For this and other reasons, it should not be eaten. Some authors list *Pholiota aurivella* as edible. Others report that this fungus contains gastrointestinal-upsetting compounds. You be the judge!

FAMILY: Strophariaceae (Cortinariaceae)

GENUS AND SPECIES: *Pholiota carbonaria* A.H. Sm. (1944) (Fig. 91)

COMMON NAME: Viscid "flammula" or burnt-ground pholiota

STRIKING FIELD CHARACTERS: On burnt ground of very recent forest fires, associated with the thalloid *Marchantia latifolia* (or possibly other *M.* species), yellowish red to orange-brown disc and yellow margin.

MACRO AND MICRO DESCRIPTION: **Cap** 1–3.5 cm broad, sticky, yellow margin, yellow-red to orange-brown disc; shiny, smooth; bell-shaped in youth becoming convex with broad umbo to broadly convex, cushion-shaped, almost plane and depressed disc in age; flesh yellow. **Gills** grayish cream, clay to grayish brown; broadly attached, crowded, thin, narrow. **Stalk** 1–4 cm long, 3–6 mm broad, apex white to pale lemon yellow, orange to reddish-tinged below and blackened at base; tough-pliant, equal to slightly tapering

Figure 91. *Pholiota carbonaria* A.H. Sm. (1944)

toward base; fibrillose base with mycelium enmeshing substrate; flesh yellowish white. Veil fibrillose and disappearing or shriveling, cinnabar red, fading to ochre and leaving some remnants on cap margin and stalk, but no ring. **Spores** are brown, elliptical, smooth, 6–7 x 4–4.5 µm.

HABITAT AND ROLE: This relatively small fungus will be found singly or scattered to gregarious in massive clusters all over a recently (previous summer) burned landscape. We once found it in great abundance west of Kantishna, where some very large forest fires burned through black spruce and larch (*Larix laricina*) muskeg. We have also seen areas of the forest floor carpeted with *Marchantia latifolia.* It was virtually everywhere underfoot, and we could not walk without stepping on fruiting bodies. If you get into this kind of an area, you are apt to also see other wood/charcoal-decomposing mushrooms such as *Pholiota fulvozonata,* with a darker tawny cap and russet veil; *P. highlandensis,* with a brown cap and yellow veil; *Hebeloma anthracophilum*; *Geopetalum carbonarium* (= *Cantharellus umbonatus* var. *carbonarious* [Alb. & Schwein.] Fr and *Faerberia carbonaria*); or various cup fungi such as *Anthracobia* spp., *Geopyxis carbonaria, Peziza* spp., and *Ascobolus* spp. Also, the morel fungi are particularly abundant on such sites the first spring after forest fires.

EDIBILITY, TASTE, AND ODOR: This fungus has been described as being edible, but of poor quality. Other authors do not even comment on its edibility. Its taste is slightly bitter, while the odor is of burnt soil or fireplace ashes.

FAMILY: Strophariaceae (Cortinariaceae)

GENUS AND SPECIES: *Pholiota squarrosoides* (Peck) Sacc. (1887) (Fig. 92)

COMMON NAME: Sharp-scale, scaly, or bristly pholiota

STRIKING FIELD CHARACTERS: Cream to yellow-brown (tawny) tacky to viscid (sticky) cap, and dry stalk; scales orange-brown; stalk apex white to creamy buff above a fibrillose ring; flesh dull greenish yellow (light olive), white below; citron to olive-yellow stalk center and yellow base; on birch.

Figure 92. *Pholiota squarrosoides* (Peck) Sacc. (1887)

MACRO AND MICRO DESCRIPTION: **Cap** 1.5–7.5 cm broad, broadly conic to convex when fully expanded; ivory, yellowish buff, cream at first to yellowish brown to orange-tan (tawny) in age; scales darker golden brown to orange-brown, pointed; tacky to viscid (sticky) in wetter weather; flesh firm, white to light, dull olive; disc often devoid of upright scales in age. **Gills** notched to broadly attached, close; white to cream, yellow-buff, dull yellow-brown to rusty brown, unequal. **Stalk** 3–10 cm long, 5–15 mm broad, equal to tapering toward caespitose base; dry; ground color white, cream, buff, to tawny; scales recurved, dull buff to walnut brown; context white on outside, olive-yellow inside, to yellow at the base. **Spores** dull reddish brown, without a germ pore, elliptical, smooth, 4–6 x 3–3.5 µm.

HABITAT AND ROLE: This lignicolous fungus is found in dense tufted clusters at the base of or in the crotches of paper birch within mixed aspen, birch, and white spruce forests. It is a suspected root pathogen or heart rot fungus as well.

EDIBILITY, TASTE, AND ODOR: Unlike its close relative *Pholiota squarrosa*, which may act as a gastrointestinal irritant, *P. squarrosoides* is reported as edible. Its mild to nutty taste, mild green corn odor, viscid (sticky) cap, and flattened cap scales in wet weather will separate this edible fungus from the dry cap and stalk of *P. squarrosa*, which also has a green tint to the gills and a strong odor of fresh green corn.

FAMILY: Cortinariaceae

GENUS AND SPECIES: *Rozites caperatus* (Pers.) P. Karst. (1879) (Fig. 93)

COMMON NAME: Chicken of the woods, gypsy, goat pholiota, or wrinkled pholiota

STRIKING FIELD CHARACTERS: Cap light yellowish tan to orange-tan, wrinkled towards margin and "frosted" or with white dustlike surface in youth, with white annulus on stalk.

MACRO AND MICRO DESCRIPTION: **Cap** 3.5–10 cm broad, oval in youth and expanding to convex, becoming broadly convex, broadly knobbed over disc (center), dry, finally wrinkled to margin and with a dusting or frosted appearance that disappears in age. Margins are turned in when young, becoming straight with expansion. Flesh thick, firm, white to buff, pleasant smelling. **Gills** notched, narrowly to broadly attached to stalk, close; dull leather tan, becoming dull rusty cinnamon with age. **Stalk** 0.6–2 cm wide, 5–15 cm long, tapering and longitudinally fibrous from a swollen (sub-bulbous) base to the apex; smooth, dry; white to dull buff to light tan. **Annulus** is membranous, whitish, and thin. **Spores** red-brown, elliptical, rough-warty, 11–15 x 7–10 µm.

Figure 93. *Rozites caperatus* (Pers.) P. Karst. (1879)

HABITAT AND ROLE: This mushroom is a suspected mycorrhizal symbiont with our dwarf birch (*Betula nana* subsp. *exilis* and *B. glandulosa*) and white spruce at or near tree line. Here it is found singly to numerous and down in amongst the ground cover in late July to mid-August, where it can be seen in quantity in some years.

EDIBILITY, TASTE, AND ODOR: This mushroom is considered edible and good, often slightly bitter when eaten raw but delicately delicious when sautéed. Taste is mild to nutty and odor is very pleasant. Since this is a close relative to members of the genus *Cortinarius*, one would want to make sure the veil is membranous and persistent rather than being cobweblike, as is typical for many Cortinarii that may cause serious gastrointestinal problems or death. Close look-alikes to *Rozites caperata* are some *Agaricus* species that have pink flesh to dark chocolate brown gills that are not attached to the stalk.

FAMILY: Strophariaceae

GENUS AND SPECIES: *Hypholoma (Naematoloma) udum* (Pers.) Kühner (1877) (Fig. 94)

COMMON NAME: Moss hypholoma

STRIKING FIELD CHARACTERS: Small, yellowish buff to light tan cap; broadly bell-shaped to mostly convex with or without a broad umbo or slightly raised disc; grayish brown to purplish brown mottled gills, thin; stalk long and pliant and with a collapsing annulus.

MACRO AND MICRO DESCRIPTION: Cap 0.6–20 mm broad, broadly bell-shaped to mostly convex, and with or without a slightly raised umbo over the disc; buff to yellow and often somewhat darker tan to light brown over moist to viscid disc; margin without partial veil remnants attached and somewhat flaring in age; flesh dirty white and thin. **Gills** thin, narrow, unequal, mottled dingy purple to grayish brown, close together. **Stalk** 7–8 cm long, 2–4 mm broad, equal to tapering slightly toward apex; smooth, hollow; buff to yellowish buff (lighter than cap), flexuous, pliant. Veil minute, membranous at first, becoming fibrillose, and finally disappearing except for a ring of purple-brown spores deposited previously onto the veil prior to its collapsing against the stalk. **Spores** purple-brown in deposit, pored, elliptical, smooth, 12–19 x 6–8 µm.

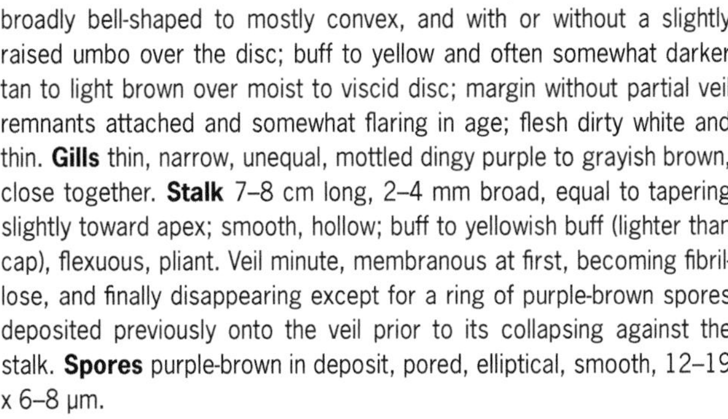

Figure 94. *Hypholoma (Naematoloma) udum* (Pers.) Kühner (1877)

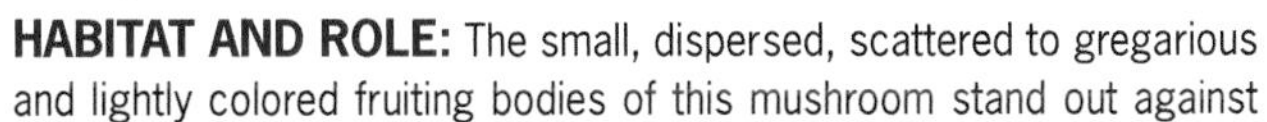

HABITAT AND ROLE: The small, dispersed, scattered to gregarious and lightly colored fruiting bodies of this mushroom stand out against the dark green moss-covered (*Polytrichum commune*, *P. juniperinum*, or *P. strictum*) forest floor under white spruce, aspen, and birch. *Hypholoma (Naematoloma) udum* is a saprobic decomposer of moss parts and other plant debris. In wetter areas, this fungus is often found on unrecognizable organic detritus in depressions, along old compacted trails, or in peaty soil disturbed by heavy equipment. It is relatively small, so keep your eyes open.

EDIBILITY, TASTE, AND ODOR: *Hypholoma (Naematoloma) udum* is too small to collect for the table, although its taste is mild to nutty and its odor earthy, most often like that of the moist humus or moss upon which the fungus resides.

FAMILY: Strophariaceae

GENUS AND SPECIES: *Stropharia semiglobata* (Batsch) Quél. (1872) (Fig. 95)

COMMON NAME: Dung, round head, or hemispheric stropharia

STRIKING FIELD CHARACTERS: Straw yellow to buff-tan, sticky to slimy, rounded cap; long, slender, and annulated stalk; light purplish gray gills.

MACRO AND MICRO DESCRIPTION: Cap 2–4 cm broad, conic, parabolic (bell-shaped but narrowed at basal margin), pulvinate (cushion-shaped), convex to almost plane; color pale cream, straw, yellowish buff to buff-tan to light yellowish brown; smooth, sticky; margin lighter; flesh whitish buff. **Gills** notched to broadly attached and sometimes slightly decurrent and breaking away from the stalk (seceding); color gray-buff when young and then becoming lavender, more or less mottled, to a light purplish gray to purple-brown and finally to purple-black; uneven, crowded at margin, much less so (subdistant) at stalk; thin and fairly narrow. **Stalk** 6–20 cm long, 3–8 mm broad; long and slender, tapering gently toward apex to equal and enlarging toward base; sticky to slimy glutinous; buff to yellow-tan and lighter than cap disc, often same color as the cap margin. Veil (partial) delicate to fragile, membranous to fibrillose, collapsing about stalk and often blackened from spore deposits on upper surface. **Spores** dark purple-brown, elliptical, smooth, 15–19 x 7.5–10 µm.

Figure 95. *Stropharia semiglobata* (Batsch) Quél. (1872)

HABITAT AND ROLE: This fungus has been found as solitary or scattered to rarely gregarious fruitings in mixed aspen–birch–white spruce forests, emanating from what looks like old bear scat and/or badly degraded moose nuggets located under deep moist moss carpets.

EDIBILITY, TASTE, AND ODOR: This fungus is reported to be poisonous, nonpoisonous, to edible but mediocre by various sources and is probably of little or no interest due to its relatively small size. At best, it is highly questionable and for that reason we suggest you leave this one for the squirrels, caribou, and bears. Its taste has been recorded as mildly bitter. Our specimens taste nutty with a slightly bitter aftertaste. Odor is mild to nondistinct.

FAMILY: Psathyrellaceae (Coprinaceae)

GENUS AND SPECIES: *Coprinus atramentarius* (Bull.) Fr. (1838) (Figs. 96–97)

COMMON NAME: Gray ink-cap, tippler's bane, or dark flaky inky cap

STRIKING FIELD CHARACTERS: Thick, fleshy, light grayish tan, radially lined, and often-splitting cap; hollow white stalk with a flaring basal "ring" from which the cap margin "detaches" in time.

Figure 96. *Coprinus atramentarius* (Bull.) Fr. (1838)

MACRO AND MICRO DESCRIPTION: Cap 1.5–5 cm broad, ovoid, conic, bell-shaped to convex; dry, smooth, striate to furrowed; gray, gray-brown, gray tan to light brown (dark tan); margin striate to furrowed to wrinkled; silky and sometimes with flattened scales; flesh thin, off-whitish gray. **Gills** white, pinkish gray to black; crowded, thin, broad, nearly free, deliquescing (turning to inky juice in time). **Stalk** 2.5–7.5 cm

Figure 97. *Coprinus atramentarius* (Bull.) Fr. (1838)

long, 6–12 mm broad, tapering from basal "ring" toward apex and again toward base; color white and smooth above "ring," whitish and fibrillose below "ring"; hollow, brittle; partial veil leaving basal "ring" on stipe. **Spores** black, elliptical, smooth, 7–12 x 4–6 μm.

HABITAT AND ROLE: A terrestrial saprobic decomposer, this fungus is rarely solitary, but more commonly occurs in clusters on lignin-rich humus within mixed white spruce–aspen forests.

EDIBILITY, TASTE, AND ODOR: *Coprinus atramentarius* has been described as edible, choice, or edible with caution. It is edible when young (when nondeliquescing) and fairly good, but is reported to contain a disulferam-like compound (Coprine) that causes vomiting if consumed along with an alcoholic beverage. Coprine reportedly reacts with alcohol and may produce reddening of ears and nose, metallic taste, lightheadedness, tachycardia (rapid heartbeats), throbbing sensations, nausea, and vomiting. Scott Chilton, a well-known pharmacologist of the University of Washington who used to study mushroom toxins, once suggested that these toxins contain potentially bad liver poisons, and should be avoided.

FAMILY: Agaricaceae (Coprinaceae)

GENUS AND SPECIES: *Coprinus comatus* (O.F. Müll.) Gray (1797) (Fig. 98)

COMMON NAME: Shaggy mane, inky cap, horsetail, inky eggs, lawyer's wig, shaggy beard, or maned agaric

STRIKING FIELD CHARACTERS: Long ovoid to tall straight cylinder-shaped; light tan to darker brown-scaled tips; large, stately, and on compact but disturbed (roadside, gravelly) soil.

MACRO AND MICRO DESCRIPTION: **Cap** 5–30 cm tall, barrel-shaped in youth to cylindrical in age, 2.5 to 6.3 cm broad and becoming bell-shaped in age, white with tan to brown apex and covered with tiers of shaggy, fibrous, recurved, or reflexed (upturned) scales. Flesh soft, white, and thin. Margin slightly incurved and with a fibrous veil in youth becoming turned out, shaggy to dissolving into an inky juice (liquefaction or deliquescence) with age. **Gills** narrowly attached (notched) to stalk, white, becoming pinkish to smoky and finally black and then dripping away into a black watery juice, densely crowded, and thin. **Spores** black in deposit. **Stalk** 5–36 cm long, 6–23 mm broad, cylindrical, equal to tapering slightly toward apex; dry, white, with an expanded to sub-bulbous base, ringed. Look for a white frilly central "cord" that attaches to the cap but runs almost to the base within the hollow stipe. Spores with a germ pore, elliptical, smooth, 10–16 (–18) x 7–9 μm.

Figure 98. *Coprinus comatus* (O.F. Müll.) Gray (1797)

HABITAT AND ROLE: These mushrooms are solitary, scattered, gregarious to densely clustered in compact groups on hard ground and grassy areas around the hotel complex, DNP&P headquarters, and along

the road system. They are known to be saprophytic (digesting dead organic litter or detritus) and are quite common during mid- to late summer (July). This mushroom has been seen pushing up through hard-packed gravel roads and even asphalt with relative ease.

EDIBILITY, TASTE, AND ODOR: These fungi are edible and very good, more so when young. You will not find them to be particularly palatable when they begin the liquefaction stages of development. The enzymes that cause their deliquescence are quite active at or near "refrigerator" temperatures (2–4°C/35–37°F), so do not wrap them neatly in waxed paper or paper bags and expect to find anything but black dripping ooze all over the inside of your refrigerator the next morning. Freezing works if the specimens are young. Once again, this fungus is not easily mistaken for any other. Taste is mild but very good and odor is very pleasant, fresh, and clean-smelling. Another mushroom the untrained eye might confuse with the shaggy mane during certain stages of development is *Coprinus atramentarius*. Shorter, stouter, grayish in color, without scales, and more bell-shaped, *C. atramentarius* should never be eaten if any alcohol is to be consumed, as they, too, contain liver poisons and should be avoided.

FAMILY: Agaricaceae (Coprinaceae)

GENUS AND SPECIES: *Coprinus micaceus* (Bull.) Fr. (1838) (Fig. 99)

COMMON NAME: Glistening inky cap, mica cap mushroom

STRIKING FIELD CHARACTERS: Ovoid to conic with a tan to yellowish brown cap with glistening (mica-like) particles, white stalk, in dense aggregate clusters on "woody" humus.

MACRO AND MICRO DESCRIPTION: Cap 1–4 cm broad, ovoid to conic when expanding; dry; buff, tan to yellowish orange to orange-brown; covered with glistening white disappearing particles (pulverulent); margin striate to furrowed to rugose (rough) and uneven; flesh stuffed, soft, white to off-white. **Gills** crowded, narrow, notched; white to black, deliquescing (liquefying) in age. **Stalk** 2.5–7.5 cm long, 3–7 mm broad, smooth, white, brittle, hollow, tapering toward apex; with many woolly scales (flocculent) to somewhat roughened (subscabrous) near base. Partial veil, if present, evanescent (withering) and of mica-like granules. **Spores** dark brown to black, elliptical, smooth, 7–11 x 4–6 μm.

Figure 99. *Coprinus micaceus* (Bull.) Fr. (1838)

HABITAT AND ROLE: Forming aggregate clusters on earth (terrestrial) and on rich woody humus of mixed alder, birch, aspen, and spruce forests. This fungus is a saprobic decomposer of disturbed and/or open areas and often around shrubs.

EDIBILITY, TASTE, AND ODOR: *Coprinus micaceus* is reported to be edible, good but not choice due to its thin flesh and watery consistency. Its flavor is also reported as good. Its taste is watery to mealy and the odor is pleasant. This fungus does not always deliquesce completely or at all and, therefore, it might even be confused (macroscopically) with the genus *Psathyrella* of this same family, which are best left alone.

FAMILY: Agaricaceae (Coprinaceae)

GENUS AND SPECIES: *Coprinus* aff. *patouillardii* Quél. (1884) (Fig. 100)

COMMON NAME: Fragile moose nugget inky cap

STRIKING FIELD CHARACTERS: Minute, translucent white cap with few gills; long, slender, and fragile stalk; on moose nuggets (i.e., coprophilous).

MACRO AND MICRO DESCRIPTION: **Cap** minute, 2–7 mm broad, ovoid, conic to plane and/or depressed in age; flesh thin, watery, fragile; plicate (pleated) and when broadly expanded almost smooth; margin crenate (wavy), pellucid-striate, "translucent" white. **Gills** are thin, narrow, spaced to distant (24–32 per cap), attached, grayish white but blackening near or along edges in age. **Stalk** 20–35 mm long, 1–2 mm wide, glistening grayish white half way to the apex, even to flexuous, minutely flocculent (woolly-scaly) and fibrillose below to base. **Spores** black in deposit, heart-shaped, thin in profile view, 6–10 (–12) x 5–7 (–10) x 3.5–5 µm.

Figure 100. *Coprinus* aff. *patouillardii* Quél. (1884)

HABITAT AND ROLE: A coprophilous (dung-inhabiting) fungus, the minute, glistening white fruiting bodies of *Coprinus patouillardii* are found only on moose nuggets, with these often buried in moist to wet mossy depressions under willows and white spruce–bordering streams.

EDIBILITY, TASTE, AND ODOR: Edibility is unknown, as the fungus is so small that it is simply not economical to attempt gathering enough to cook, and then we are not sure how! Taste is unknown, but the odor is often faintly fishy, probably from the wet mossy and coprophilic habitat it occupies. We suggest that you not snack on this species, as you would probably get more nutritional value from the "substrate" on which it grows than from the fungus itself!

FAMILY: Bolbitiaceae (Coprinaceae)

GENUS AND SPECIES: *Panaeolus* aff. *acuminatus* (Schaeff.) Quél. (1874) (Fig. 101)

COMMON NAME: Moose dung panaeolus

STRIKING FIELD CHARACTERS: Dark, dull black, purple to gray-brown bell-shaped cap; thin dark stalk; mottled gray to purple-brown gills; on moose dung.

MACRO AND MICRO DESCRIPTION: **Cap** 3–5 mm broad, bell-shaped; smooth, moist, subhygrophanous; black, purple-brown to grayish brown, dull much lighter gray-brown when very young; appearing almost zonate or banded; flesh dull gray-brown, thin; margin often lighter whitish brown to gray-brown in extreme youth, becoming almost transparent to translucent-striate to subfurrowed in age. **Gills** light gray-brown to purple-brown to almost black; mottled and notched. **Stalk** 12–75 mm long, 1–3 mm broad; smooth; light grayish brown to purple-brown; equal, brittle; base often clothed in white

Figure 101. *Panaeolus* aff. *acuminatus* (Schaeff.) Quél. (1874)

to gray mycelium. **Spores** black, pored to truncate, 12–15 x 8.5–11 µm.

HABITAT AND ROLE: This fungus is either too small or too darkly colored to be conspicuous. Nevertheless, it is always found on old piles of moose dung along rivers and creeks, nestled neatly or deeply down in the moss or detritus of riparian willow (*Salix alaxensis*) and white spruce forests. It is a decomposer of the mostly woody *Salix* content of the moist dung. In the alpine zone, we find it on vole dung. In the Arctic, it is found on lemming dung.

EDIBILITY, TASTE, AND ODOR: The edibility of this fungus is not reported, perhaps not even known. Considering its substrate, few would venture to eat it, if you could even find enough for a meal. Several close relatives, *Panaeolus foenisecii* and others, are potentially poisonous.

Boletinoid (Soft, Fleshy Poroid) Fungi

The bolete fungi (see Fig. 16b) include some of the largest mushrooms likely to be encountered in the Alaska interior. All have a fleshy hymenophore consisting of many tubes with pore openings. This pore-filled hymenium (or sponge) is found on the underside of the cap, as in the agaric fungi. Boletes are "fleshy," often brightly colored, and become riddled with the brownish entry holes of fly larvae, which are easily seen if you peel the hymenium away from the cap. All boletes found in Alaska are members of two families, the Boletaceae and Suillaceae, and the majority of them belong to five genera, *Boletus*, *Chalciporus*, *Leccinum*, *Suillus*, and *Xerocomus*.

All the interior Alaska boletes except one, *Leccinum atrostipitatum*, are edible, although not necessarily palatable. They may become slimy when cooked, ill-textured for the palate, too peppery, or may simply not meet your taste expectations. Avoid those infected by insect larvae and fruiting bodies too old to consume safely. Bacteria and their metabolic waste products can make one far sicker than the original mushroom tissue itself. Be selective! Prepare for your camp table only those that are young and freshly picked, but *always* cook them.

FAMILY: Boletaceae

GENUS AND SPECIES: *Boletus edulis* Bull. (1782) (Fig. 102)

COMMON NAME: Edible bolete, steinpiltz, king bolete, porcini, or cep

STRIKING FIELD CHARACTERS: Light to dark brown, bumpy to almost greasy-looking cap; netted over pale buff stalk, particularly at apex; firm flesh; pores white to yellow-green.

Figure 102. *Boletus edulis* Bull. (1782)

MACRO AND MICRO DESCRIPTION: Cap 4–30 cm broad, hemispherical, convex to broadly so and then tabletop flattened in extreme age; white at margin otherwise biscuit-colored to yellow, cinnamon brown to reddish brown; surface mostly dry but in wet weather it can be sticky; bumpy to pitted to cracked in dry weather; flesh is thick, firm, white, not "bluing" when exposed or bruised. Pores white to whitish yellow at first and then becoming yellow, olive-yellow, or tan in extreme age; small, circular, long tubular and can be coerced to "peel" away from the cap. **Stalk** 4–10 cm long, 2–6 cm broad, bulbous, and less so when expanded or extended upward; firm, solid, white, pallid to buff-colored, covered by a netlike white reticulum most easily seen at the apex. **Spores** olive-brown, spindle-shaped to elliptic, smooth, 13–19 x 4–7 µm.

HABITAT AND ROLE: This fungus serves a very important fungus–plant root (mycorrhizal) relationship with our deciduous trees, especially birch and aspen, where you are most likely to find it growing singly to scattered, rarely in groups or troops, over the forest floor. It is most apt to be found during July in the Alaska interior.

EDIBILITY, TASTE, AND ODOR: This fungus will be seen and sold dried in specialty (gourmet) and health food stores. It is a delicious edible and reconstitutes nicely after having been dried (preserved). It has a pleasant to nutty taste and fresh, pleasant odor. Look for this one. Sautéed over the campfire with butter and a little salt and garlic, *Boletus edulis* makes a great appetizer or additive to a thick juicy steak and is often used in dehydrated mushroom soups, especially in Europe.

FAMILY: Boletaceae

GENUS AND SPECIES: *Xerocomus subtomentosus* (L.) Fr. (1821) (Fig. 103)
= *Boletus subtomentosus* L. (1753)

COMMON NAME: Brown and yellow bolete

STRIKING FIELD CHARACTERS: Dry, olive-yellow to dark olive-brown, and velvety-tomentose (soft, short-haired surface) cap, generally without cracks, bright golden-yellow pores and a buff to creamy tan stalk of fairly equal proportions throughout.

MACRO AND MICRO DESCRIPTION: Cap 4–10 cm broad, broadly convex to plane; dry and minutely velvety (subtomentose) cap that in age may look and feel smooth; yellow-brown to drab olive-brown to dull brown to cinnamon brown; flesh white, off-white (pallid) cream to straw yellow. Pores are large (1–3 mm), dull to bright chrome yellow, staining blue-green, but weakly so. **Stalk** 4–10 cm long, 1–2 cm wide; when short, clavate but with extension equal to and tapering toward apex; smooth to scurfy, but reticulations are present at the apex; flesh firm and same color as

Figure 103. *Xerocomus subtomentosus* (L.) Fr. (1821)

cap flesh; outer surface off-white, buff, or tan with orange tinges. **Spores** olive-brown, spindle-shaped to elliptic, smooth, 10–16 x 3.5–5 µm.

HABITAT AND ROLE: This mushroom can be found singly to widely scattered but closely associated with dwarf and paper birch and shrub tundra members of the Ericaceae (evergreen shrubs and ground cover). It most certainly is a mycorrhizal associate of these plants.

EDIBILITY, TASTE, AND ODOR: Albeit edible, for those with a delicate palate this fungus is reported as not being particularly good to eat. Try it and decide for yourself. If you prefer bland food, this may just be the one! The odor is pleasant, certainly not obnoxious, but again, sort of bland.

FAMILY: Boletaceae

GENUS AND SPECIES: *Leccinum aurantiacum* (Bull.) Gray (1821) (Fig. 104)

COMMON NAME: Trembling or quaking aspen bolete

STRIKING FIELD CHARACTERS: With quaking aspen, *Populus tremuloides*, orange-red to orange-brown cap; stalk with short, stiff, red-brown projections (scabers); white to off-white pores; and flesh turning lilac, lilac-gray then slowly to smoky black.

MACRO AND MICRO DESCRIPTION: Cap 2–15 cm broad, convex to broadly so, nearly plane and slightly depressed in age (water will puddle), dry to tacky when wet, bright orange to rusty red to reddish brown; margin with remnant patches of marginal veil tissue; flesh dull off-whitish, firm, stains wine reddish to lilac with gray tints and then slowly blackens. Pores fairly deep (1–1.5 cm deep), round openings, small. **Stalk** 5–15 cm long, 1.5–3 cm broad, enlarging toward base, ground tissue white, but scabrous, the projections brown, dark brown to almost black. **Spores** yellow-brown to brown, spindle-shaped to elongate-elliptic, smooth, 13–18 x 3.5–5 µm.

Figure 104. *Leccinum aurantiacum* (Bull.) Gray (1821)

HABITAT AND ROLE: This stately mushroom is mycorrhizal with its host, the quaking or trembling aspen, which is relatively common in interior Alaska and particularly so in areas of recent (twenty-five to fifty years) forest fires. When found, will be single, scattered, to gregarious depending on the season and fruiting pattern. It can certainly be abundant some years!

EDIBILITY, TASTE, AND ODOR: Both taste and odor are mild, but good as also reported by several authors. In the Lower 48 states, and because of its staining reaction, some who do not know this fungus comfortably may feel less inclined to consume it. Do not hesitate with material from the Alaska interior. But just remember, our northern flies love them too.

FAMILY: Boletaceae

GENUS AND SPECIES: *Leccinum atrostipitatum* A.H. Sm., Thiers & Watling (1966) (Fig. 105a–b)

COMMON NAME: Trembling or quaking aspen bolete

STRIKING FIELD CHARACTERS: A *Leccinum aurantiacum* look-alike; dry orange, rusty orange, reddish orange, orange-brown to cinnamon brown, with marginal tissue flaps staining lilac to lilac-gray, purplish gray to smoky gray with time, with brown scabers.

MACRO AND MICRO DESCRIPTION: Cap 3–12 cm broad, round to convex, becoming broadly so to nearly plane; dry to tacky when wet, smooth to fibrillose, with the fibers separating with expansion; color from light to dark orange, rusty orange, reddish orange, cinnamon to brown; marginal tissue flaps often gone in age; robust, flesh white but staining lilac to lilac-gray to purple-gray to smoky gray. Pores dull white, greenish buff, gray to dull yellowish buff in age, not or slow staining to lavender or purplish red (vinaceous). **Stalk** 4–10 cm long, 1–2 cm broad, equal, dry, solid, off-white, scabers initially whitish brown, then light reddish orange to brown and finally dark chocolate brown to almost blackish brown in age; base often staining a greenish blue. **Spores** subfusoid, smooth, 13–17 x 4–5 µm.

Figure 105a–b. *Leccinum atrostipitatum* A.H. Sm., Thiers & Watling (1966)

HABITAT AND ROLE: This mycorrhizal fungus is associated with aspen just like *Leccinum aurantiacum*, but its staining reactions and other subtle microscopic characters separate the two species. **Fruiting bodies** of *L. insigne* occur scattered to gregarious, and are fairly common to often abundant during July at such places in the DNP&P as near the park hotel or at Wonder Lake.

EDIBILITY, TASTE, AND ODOR: These stately boletes are not edible, should not be eaten, and have proven to cause severe gastrointestinal discomfort. They cook up "dirty" (that is, stain gray-black) and even look inedible. Taste and odor are fairly mild, but don't let that lead you astray. Folks who have eaten this stately Alaska mushroom raw or with a favorite libation have been particularly vulnerable to its ill repute!

FAMILY: Boletaceae

GENUS AND SPECIES: *Leccinum rotundifoliae* (Singer) A.H. Sm., Thiers & Watling (1967) (Fig. 106)

COMMON NAME: Boggy birch bolete

STRIKING FIELD CHARACTERS: A white to buff *Leccinum scabrum* look-alike; pale white, yellowish white, yellowish gray with white to graying stalk.

MACRO AND MICRO DESCRIPTION: Cap 1.5–6 cm broad, convex to broadly so; smooth dry to tacky when wet; color white at margin to patchy, buff, faint pinkish tones to yellowish white, yellowish buff, yellowish tan to yellowish gray, "staining" pinkish brown where pressed against surrounding dwarf birch vegetation but otherwise not staining;

Figure 106. *Leccinum rotundifoliae* (Singer) A.H. Sm., Thiers & Watling (1967)

flesh white. Pores white to dull white, aging to buff and spotty tan; tube pores small, circular, recessed at stalk apex; not staining unless browning around insect entry holes or galleries. **Stalk** white to slightly dirty (graying) white; scabers white, buff, to light tan; 0.5–1.5 cm broad, equal to tapering at apex, to flexuous. **Spores** narrowly subfusoid to elongate-sub-ovate, inequilateral, pale ochraceous to brownish in deposit, smooth, (11.5-) 14–20 (–25) X 4.5–7.5 (–8) µm.

HABITAT AND ROLE: Single, scattered, or rarely gregarious, this fungus is commonly found along with *Leccinum scabrum* under dwarf birches. But where *L. scabrum* tends to be on low-angled hillsides and upper slopes, *L. rotundifoliae* "replaces" it in lower, wetter, flatter, depressed birch fens. This fungus is a mycorrhizal associate of the dwarf birch species with which it is found.

EDIBILITY, TASTE, AND ODOR: Taste and odor of *Leccinum rotundifoliae* are mild. It is also edible and good but goes by quickly as the tissues tend to be less compact, softer, and therefore seemingly more vulnerable to the ubiquitous fly pests. Watch out for the protein you may not want with your meal.

FAMILY: Boletaceae

GENUS AND SPECIES: *Leccinum scabrum* (Bull.) Gray (1821) (Fig. 107)

COMMON NAME: Birch bolete

STRIKING FIELD CHARACTERS: Tan to dull brown cap, dry, tacky when wet, no marginal tissue flaps, flesh white and unchanging or faint pink tints in stalk, with paper birch or the dwarf birches.

MACRO AND MICRO DESCRIPTION: **Cap** 3–8 (–10) cm broad, convex to broadly so and becoming nearly plane in extreme age with a depressed disc; smooth, dry to tacky in wet weather, whitish buff at margin to tan, dingy yellow-brown to dull brown; margin without tissue flaps; flesh tissue white, not staining or only very slight pinkish tone in stem tissue. Pores dull white, whitish brown, and becoming brown in age and upon drying; no bluing stains, but light tan to brown stains may result from handling. **Stalk** 6–10 cm long, 0.5–2 cm broad, tapering toward apex and sunken pores; white, buff, ochre, to tan with same-colored scabers that darken with age; staining green at or near base. **Spores** brown, spindle-shaped to long elliptical, smooth, 14–20 x 5–7 µm.

Figure 107. *Leccinum scabrum* (Bull.) Gray (1821)

HABITAT AND ROLE: Here is another mycorrhizal fungus, but this time with paper and dwarf birch. This is a variable fungus and can be almost sooty brown in youth. It is quite common and occurs as solitary to gregarious (rarely in clumps) fruitings in deciduous forests, mixed forests, or upland shrub treeline tundra.

EDIBILITY, TASTE, AND ODOR: Taste and odor are mild. This is a good edible in the north, but once again, the bugs will compete and usually win. It is hard to find good, firm, young specimens for the table that the proverbial maggots have not already discovered. But then, a little protein with your meal may not hurt!

FAMILY: Suillaceae (Boletaceae)

GENUS AND SPECIES: *Suillus cavipes* (Opat.) A.H. Sm. & Thiers (1964) (Fig. 108)
= *Boletinus cavipes* (Opat.) Kalchbr. (1867)

COMMON NAME: Hollow foot, hollow-stiped bolete, or mock oyster

STRIKING FIELD CHARACTERS: Fruiting in muskeg bogs with larch and black spruce; dry red, streaked cap, bright yellow-orange and large pores; hollow annulate (ringed) stalk.

MACRO AND MICRO DESCRIPTION: **Cap** 3–10 cm broad, convex to broadly so, umbonate, expanding to plane, uplifted and depressed with raised disc; surface dry (not tacky when wet) red fibrils and densely fibrillose, leathery texture, margin often lighter, white fibrillose in youth, yellowish red, tawny orange-red, to red-brown in age and uplifted; flesh tissue white to yellow-orange where damaged but not staining green, blue, violet, purple, or gray when bruised. Pores radiately arranged, large, angular to elongated; pale yellow to greenish yellow, finally golden yellow and with reddish orange highlights; not bruising green, blue, violet, purple or gray; tubes "decurrent" at least in youth, less so in age. **Stalk** 3–8 cm long, 0.5–2cm broad, swollen toward base in youth, equal in age; dry; white to greenish yellow above red ring, same color as the cap below to a white base; hollow to base; flesh tissue not staining green or blue. Veil fibrillose; white cottony on margin, forming a thin ring, the latter white to pink on the stalk. **Spores** dark olive-brown, elliptical to spindle-shaped, smooth, 7–10 x 3.5–4 μm.

Figure 108. *Suillus cavipes* (Opat.) A.H. Sm. & Thiers (1964)

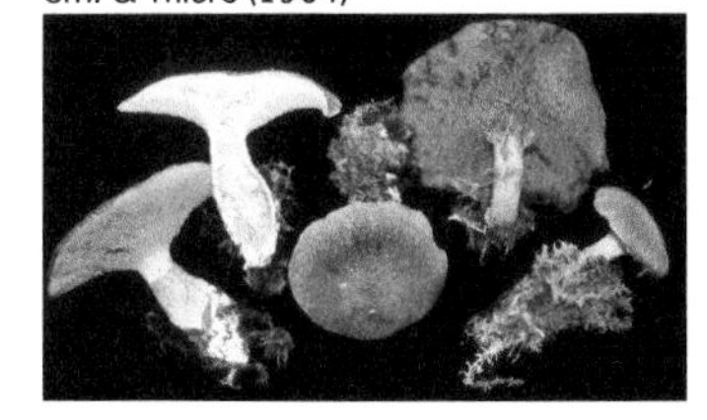

HABITAT AND ROLE: *Suillus cavipes* is found single to scattered around muskeg hummocks in moss under larch and black spruce in July. This species is a mycorrhizal associate of larch.

EDIBILITY, TASTE, AND ODOR: The taste and odor of *Suillus cavipes* are mild. It is an edible mushroom, but cooks up to a soft, almost mushy to slimy texture. Removing the tube layer helps, but hey, that is where the nutrients are; hence, the name "mock oyster" because it slides right down the gullet!

Polyporoid (Bracket- and Wood-Inhabiting) Fungi

With very few exceptions, the polypore fungi (see Fig. 16c) are tough, leathery, corky or woody, and may persist (even though not actively growing) for several years. These fungi are found on wood (living trees, dead snags, fallen logs, stumps, or buried wood) that is in some state of being decomposed (rotting away), by these and other fungi and bacteria.

Like the boletes, polypores may have a stalk, although this is uncommon. Most members of this group form brackets or shelves (perpendicular to the substrate on an upright tree, snag, stump, or downed log). The hymenium, like that of the boletes, is composed of many, but generally much smaller, pores. This pore surface (the hymenophore) is, with only a few exceptions, tough, woody, leathery, corky, and not fleshy. As a result, it really should not be too difficult to separate the wood-rotting polypore fungi from the stalked and fleshy boletes. The pore layer is not readily separable from the cap—as is found in most boletes. Unlike most other fleshy fungi, these may be annuals or perennials in growth habit.

None of the polypore fungi found in the Alaska interior would be considered edible. They are not necessarily poisonous, but simply not palatable as they are just too tough to chew.

FAMILY: Atheliaceae

GENUS AND SPECIES: *Plicatura nivea* (Sommerf.) P. Karst. (1889) (Fig. 109)

COMMON NAME: Snow-white crust

STRIKING FIELD CHARACTERS: White, flat fungus on hardwoods.

MACRO AND MICRO DESCRIPTION: **Fruiting body** flat on branch or slightly turned out, up to 10 by 3 cm; upper surface pale tan, visible surface white, often cracked, smooth or folded, spongy, soft, up to about 5 mm thick. **Spores** white, cylindrical to allantoid, smooth, 4–5 (–6.5) x 1–1.5 (–2) µm.

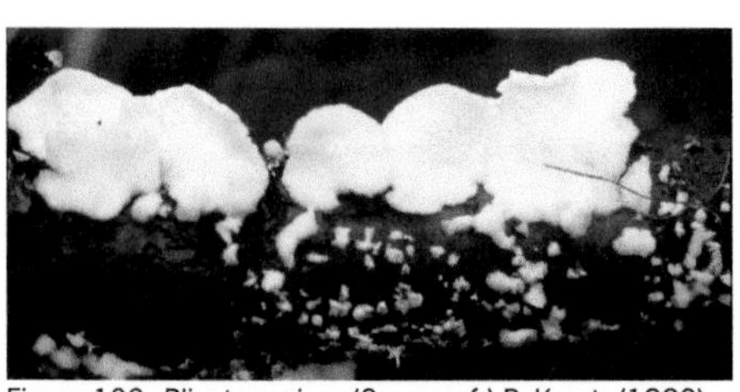
Figure 109. *Plicatura nivea* (Sommerf.) P. Karst. (1889)

HABITAT AND ROLE: This attractive white fungus causes a white rot of small branches of alder or rarely birch, balsam poplar, and willow.

EDIBILITY, TASTE, AND ODOR: *Plicatura nivea* is too small and insignificant to eat, but tempting!

FAMILY: Peniophoraceae (Corticiaceae)

GENUS AND SPECIES: *Gloeopeniophora aurantiaca* Höhn. (1908) (Fig. 110)

COMMON NAME: Painted crust

Figure 110. *Gloeopeniophora aurantiaca* Höhn. (1908)

STRIKING FIELD CHARACTERS: Flat orange paintlike crust, most often found on dead alder.

MACRO AND MICRO DESCRIPTION: Fruiting body thin and flat on branches, looking like paint, at first around lenticels (pore-like slits or holes in bark) and other holes, on undersides of branches, light or dark orange red. **Spores** salmon-colored, ellipsoid, smooth, 14–16 x 8–10 µm.

HABITAT AND ROLE: This species causes decay (white rot) of alder. Several species of *Gloeopeniophora* are similar in appearance, but *G. aurantiaca* is found only on alder. *G. incrassata*, a closely related species, is found on poplar and other deciduous trees.

EDIBILITY, TASTE, AND ODOR: All of these species are too thin and insignificant to be considered edible.

FAMILY: Ganodermataceae

GENUS AND SPECIES

= *Ganoderma applanatum* (Pers.) Pat. (1887) (Fig. 111)

COMMON NAME: Artist's conk

STRIKING FIELD CHARACTERS: Dark gray conk, white pore surface that easily bruises brown.

Figure 111. *Ganoderma applanatum* (Pers.) Pat. (1887)

MACRO AND MICRO DESCRIPTION: Fruiting bodies perennial, sessile, woody, flattened to quite irregular in shape; upper or outer surface gray to black, often with layer of brown dust (spores) present; margin narrow, white; pore surface white, bruising brown when touched, darker in age, pores circular, 4–6 per mm, tubes layered; interior tissues purple-brown. **Spores** brown to dull red-brown, broad elliptic, ovoid, truncate, thick-walled, spiny, (8–) 9–12 x 6.5–8 µm.

HABITAT AND ROLE: This species occurs on dead or living hardwoods. It is particularly common on living aspen, where it causes a white-mottled root and butt rot, but may occasionally be found on conifers.

EDIBILITY, TASTE, AND ODOR: *Ganoderma applanatum* is too hard to be edible, but makes a wonderful canvas for artists!

FAMILY: Hymenochaetaceae

GENUS AND SPECIES: *Phellinus igniarius* (L.) Quél. (1886) (Fig. 112)

COMMON NAME: Flecked-flesh polypore

STRIKING FIELD CHARACTERS: Heavy black conk with brown pore surface.

Figure 112. *Phellinus igniarius* (L.) Quél. (1886)

MACRO AND MICRO DESCRIPTION: Fruiting body perennial, single, usually sessile, hoof-shaped or flattened; upper surface usually black, becoming purple-brown; pores circular, 5–6 per mm, interior dark reddish brown, zonate, woody, off-white and brown where the fungus is in contact with the wood. **Spores** dull white in deposit (hyaline), ovoid to subglobose, smooth, 5–6.5 x 4.5–6 µm.

HABITAT AND ROLE: *Phellinus igniarius* is found on living hardwoods, mostly birch, *Betula neoalaskana*, where it causes a uniform white heart rot, surviving the death of the tree to cause decay of the dead wood.

EDIBILITY, TASTE, AND ODOR: Because of its hard woody texture, *Phellinus igniarius* is not edible, but is "ashed" and mixed with chewing tobacco by western Alaska Natives.

FAMILY: Hymenochaetaceae

GENUS AND SPECIES: *Phellinus pini* (Brot.) Bondartsev & Singer (1941) (Fig. 113)

COMMON NAME: Branch stub polypore

STRIKING FIELD CHARACTERS: Yellowish brown conk on stubs of old conifer branches.

Figure 113. *Phellinus pini* (Brot.) Bondartsev & Singer (1941)

MACRO AND MICRO DESCRIPTION: Fruiting bodies sessile or somewhat turned out from the host, solitary, or in layers; upper surface reddish brown to black; smooth or rough with ridges and zones; margin reddish brown, sometimes yellowish brown; pore surface yellowish brown, the pores circular to angular, 2–3 per mm; interior reddish brown or yellowish brown, tough, often with thin black lines. **Spores** brown, round, smooth, 4–6 x 3.5–5 µm.

HABITAT AND ROLE: The branch stub polypore causes a white pocket rot of the heartwood of living conifers.

EDIBILITY, TASTE, AND ODOR: *Phellinus pini*, which typically fruits at branch stubs, is too tough and woody to be edible.

FAMILY: Hymenochaetaceae

GENUS AND SPECIES: *Phellinus tremulae* (Bondartsev) Bondartsev & P.N. Borisov (1953) (Fig. 114)

COMMON NAME: Black birch conk (even though it grows on aspen in our forests).

STRIKING FIELD CHARACTERS: Black conk with brown pore surface, on aspen.

Figure 114. *Phellinus tremulae* (Bondartsev) Bondartsev & P.N. Borisov (1953)

MACRO AND MICRO DESCRIPTION: **Fruiting bodies** perennial, single, sessile, at branch scars; triangular in longitudinal section; with white and brown mottled core of tissue at the decayed branch stub; upper surface black, crustlike and cracked; pore surface purple-brown, the pores circular, 5–7 per mm; interior dark reddish brown. **Spores** hyaline (colorless), smooth, subglobose, thick-walled, 4.5–5 x 4–4.5 µm.

HABITAT AND ROLE: This species is very similar to *Phellinus igniarius,* but is restricted to aspen, where it causes white rot decay.

EDIBILITY, TASTE, AND ODOR: As in the case for *Phellinus igniarius*, this fungus is also too hard and tough in texture to eat.

FAMILY: Albatrellaceae (Polyporaceae)

GENUS AND SPECIES: *Albatrellus ovinus* (Schaeff.) Kotl. & Pouzar (1957) (Fig. 115)

COMMON NAME: Sheep polypore

STRIKING FIELD CHARACTERS: This fungus can be distinguished from species of boletes, which it resembles, by the pores that extend down the stalk and the hymenophore is not separable from the rest of the tissue of the stalk and cap.

Figure 115. *Albatrellus ovinus* (Schaeff.) Kotl. and Pouzar (1957)

MACRO AND MICRO DESCRIPTION: **Fruiting bodies** annual, with central to lateral stalk, usually single, cream-colored to pale brown, slightly velvety stalk, circular to subcircular, upper surface cream-colored to tan, smooth, later with cracks of pale yellow coloration, pore surface pale yellow, pores circular near the margin, angular toward the stem, 3–5 per mm, flesh cream colored to pinkish cream, fleshy, with a dark layer above the tubes. **Spores** white, ovoid, subglobose, to short-elliptic, smooth, hyaline, thick-walled, 4–5 x 3–3.5 µm.

HABITAT AND ROLE: This species is a mycorrhizal associate (terrestrial) of spruce in the Alaska interior. As such, it occurs on the forest floor rather than on wood and attaches to the spruce roots below the soil surface. It is not a wood rot fungus.

EDIBILITY, TASTE, AND ODOR: Because of its dense texture, this species would not be considered edible.

FAMILY: Polyporaceae

GENUS AND SPECIES: *Fomes fomentarius* (L.) J.J. Kickx (1867) (Fig. 116)

COMMON NAME: Tinder polypore

STRIKING FIELD CHARACTERS: Gray conk, brown pore surface, a white mottled heart rot of deciduous trees, and usually on birch.

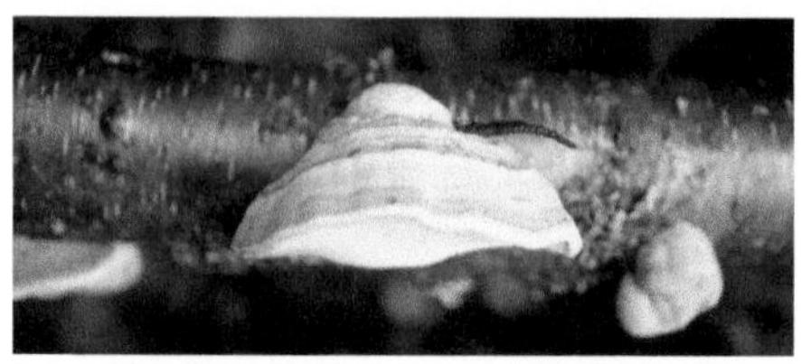

Figure 116. *Fomes fomentarius* (L.) J.J. Kickx (1867)

MACRO AND MICRO DESCRIPTION: **Fruiting bodies** perennial, sessile, hoof-shaped, woody, upper surface hard, with a smooth crust, older areas gray, margin light brown, rolled over edge; pore surface pale brown, pores 4–5 per mm; flesh yellowish brown with several tube layers comprising most of the interior. **Spores** hyaline, cylindric, smooth, 12–18 x 4–7 µm.

HABITAT AND ROLE: This species is commonly found on living and dead birch and alder.

EDIBILITY, TASTE, AND ODOR: The fruiting bodies of *Fomes fomentarius* are punky-hard and thus inedible.

FAMILY: Fomitopsidaceae (Polyporaceae)

GENUS AND SPECIES: *Fomitopsis pinicola* (Sw.) P. Karst. (1881) (Fig. 117)

COMMON NAME: Red-belted polypore, bear's bread.

Figure 117. *Fomitopsis pinicola* (Sw.) P. Karst. (1881)

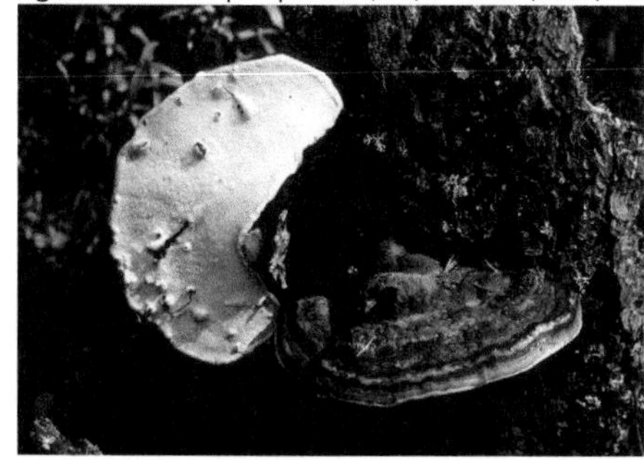

STRIKING FIELD CHARACTERS: Narrow shelf with red (orange) margin, on spruce.

MACRO AND MICRO DESCRIPTION: **Fruiting bodies** perennial, usually sessile, woody, flat, smooth; upper surface of young caps with a sticky reddish brown surface, older parts gray to black; pore surface cream colored, pores circular, 5–6 per mm; interior cream to buff, rubbery to woody, tubes in layers. **Spores** hyaline, cylindric-ellipsoid, smooth, 6–9 x 3.5–4.5 µm.

HABITAT AND ROLE: *Fomitopsis pinicola* occurs on both living and dead spruce in interior Alaska and causes a brown cubical rot of many conifers, live or dead.

EDIBILITY, TASTE, AND ODOR: This fungus is tough and rubbery and thus inedible. It also possesses a rather strong pungent/acidic odor that is not particularly appealing.

FAMILY: Fomitopsidaceae (Polyporaceae)

GENUS AND SPECIES: *Piptoporus betulinus* (Bull.) P. Karst. (1881) (Fig. 118)

COMMON NAME: Birch polypore

STRIKING FIELD CHARACTERS: Whitish cream to light tan, smooth leathery conk, on topless dead ("widowmaker") birch.

Figure 118. *Piptoporus betulinus* (Bull.) P. Karst. (1881)

MACRO AND MICRO DESCRIPTION: Fruiting bodies are annual, semicircular, sessile or stalked (stipitate) and if stalked, then with a short stalk; often pendent, solitary, cracks in age; margin white, usually rolled down over the edge of pore surface; pore surface white, yellowish brown in age, the pores circular to angular, 3–5 per mm, pore surface often appearing brushlike in old specimens; flesh white, tough when fresh. **Spores** hyaline, cylindrical, slightly allantoid, smooth, 5–6 x 1.5–2 μm.

HABITAT AND ROLE: This species causes a brown cubical rot of dead birch sapwood.

EDIBILITY, TASTE, AND ODOR: The texture of *Piptoporus betulinus* is not undesirable in young specimens, but the taste is reported to be quite bitter. This and other birch polypores are known to be dried, ground up, and the powder mixed with smoking tobacco by Native peoples for its analgesic effects.

FAMILY: Polyporaceae

GENUS AND SPECIES: *Polyporus brumalis* (Pers.) Fr. (1818) (Fig. 119)

Figure 119. *Polyporus brumalis* (Pers.) Fr. (1818)

COMMON NAME: Winter polypore

STRIKING FIELD CHARACTERS: Tough, fibrous, brown-pored fungus.

MACRO AND MICRO DESCRIPTION: Fruiting bodies solitary, annual, with central stalk, light to dark brown, cap up to 6 cm diam; upper surface light brown to dark brown, with short, dark hairs; margin light brown to dark brown, often finely fringed; pore surface white to off-white, smooth, pores angular, 3–4 per mm; flesh white, up to 3 mm thick. **Spores** hyaline, cylindric, slightly curved, smooth, 6–7.5 x 2–2.5 μm.

HABITAT AND ROLE: The winter polypore causes a white rot decay of hardwoods (usually stems of smaller diameters).

EDIBILITY, TASTE AND ODOR: The fruiting bodies of *Polyporus brumalis* are a little too tough to be considered edible.

FAMILY: Schizoporaceae (Polyporaceae)

GENUS AND SPECIES: *Schizopora paradoxa* (Schrad.) Donk (1967) (Fig. 120)

Figure 120. *Schizopora paradoxa* (Schrad.) Donk (1967)

COMMON NAME: Split-pored polypore

STRIKING FIELD CHARACTERS: Flat, on wood, white to cream-colored, toothy "pore" surface.

MACRO AND MICRO DESCRIPTION: Fruiting bodies flat on wood, often extensive, tough, white to cream-colored, darkening in age, up to 5 mm thick;

toothed pore surface, tube size variable, irregular to labyrinthine (maze-like); margin similar. **Spores** hyaline, ellipsoid, smooth, thin-walled, 5–7 x 3.5–4 μm.

HABITAT AND ROLE: This thin vesupirate fungus causes the white rot decay of small hardwood branches.

EDIBILITY, TASTE, AND ODOR: Because the fruiting bodies of *Schizopora paradoxa* are so thin and somewhat tough, they are not considered edible.

FAMILY: Polyporaceae

GENUS AND SPECIES: *Trametes suaveolens* (L.) Fr. (1838) (Fig. 121)

COMMON NAME: Anise-odor polypore

STRIKING FIELD CHARACTERS: White, thick, tough, with odor like that of anise, and on wood.

Figure 121. *Trametes suaveolens* (L.) Fr. (1838)

MACRO AND MICRO DESCRIPTION: Fruiting bodies variable sized, but generally smaller, annual, sessile, with an anise odor when fresh, solitary, semicircular; outer surface white to cream-colored, nearly smooth, margin rounded; pore surface white to cream-colored, pores circular to angular, 2–3 per mm; flesh white to cream-colored, tough, zoned, up to 5 mm thick; tubes up to 1 cm long. **Spores** hyaline, cylindric, smooth, thin-walled, 9–12 x 4–4.5 μm.

HABITAT AND ROLE: This attractive species causes a white heart rot of living hardwoods, especially willow. *Trametes suaveolens* also occasionally occurs on stumps and conifers.

EDIBILITY, TASTE, AND ODOR: In spite of its pleasant odor, *Trametes suaveolens* is just too tough to eat. It may make a good tea!

FAMILY: Polyporaceae

GENUS AND SPECIES: *Trichaptum biforme* (Fr.) Ryvarden (1972) (Fig. 122)

Figure 122. *Trichaptum biforme* (Fr.) Ryvarden (1972)

COMMON NAME: Violet-toothed polypore

STRIKING FIELD CHARACTERS: In many layers, on wood, rose-colored to purple pore surface or at least over the margin.

MACRO AND MICRO DESCRIPTION: Fruiting bodies annual, sessile; caps solitary or in layers, semicircular to flaplike; velvety to smooth, zoned; margin thin; pore surface purple to violet or rose colored, later pale tan, becoming toothy, pores angular, 3–5 per mm, splitting to form spines. **Spores** hyaline, cylindric, slightly curved, smooth, 6–7.5 x 2.5–3 µm.

HABITAT AND ROLE: This species causes white rot of hardwoods.

EDIBILITY, TASTE, AND ODOR: *Trichaptum abietinum*, a species nearly identical in appearance, occurs on conifers. Both species are too tough and dry to be eaten.

Toothed (Spiny or Hedgehog) Fungi

Thus far, there have been only a few species of toothed fungi found in the Alaska interior. As the common name suggests, the hymenium in this group is formed on the outside of pendent "teeth" or spines hanging down underneath the cap. The family, Hydnaceae, is represented in the Alaska interior by only a few genera, only five of which (*Hericium*, *Hydnellum*, *Hydnum*, *Phellodon*, and *Sarcodon*) are herein considered; one of these (*Hydnum*) is commonly seen around the park hotel when fruiting. The fruiting body of a tooth fungus (see Fig. 16d) is often mushroom-shaped (albeit usually tougher and/or with colored exadates) and has numerous small spines or teeth pointed downward. It is on these teeth (hymenophore) that the basidiospores are borne and from which they are forcibly discharged and drift away in air currents.

Of the Alaska interior's toothed fungi, only the white *Hericium* spp. should be considered edible. All others are tough-fleshy, not palatable, bitter, or may produce gastrointestinal anomalies. It is simply best not to attempt eating members of this group from the Alaska interior.

FAMILY: Bankeraceae (Hydnaceae)

GENUS AND SPECIES: *Hydnellum suaveolens* (Scop.) P. Karst. (1880) (Fig. 123)

Figure 123. *Hydnellum suaveolens* (Scop.) P. Karst. (1880)

COMMON NAME: Sweet blue tooth (but very bitter tasting)

STRIKING FIELD CHARACTERS: Fruiting bodies somewhat funnel-shaped, with a slightly rounded to flattened head, fuzzy, interior "zoned" and bluish purple near base.

MACRO AND MICRO DESCRIPTION: Cap of fruiting bodies up to 4–15 cm diameter, convex to flattened, surface fuzzy, soft, uneven (but may develop small cap with spines at the center), dirty white, changing slowly to brown at center and gradually towards the white margin; margin with broad sterile band, interior fibrous, zoned, with violet, brightest in stem, odor strongly fragrant of anise (licorice) or almond. Stem 3–5 by 1–2.5 cm, woody, interior zoned with dark blue bands, surface fuzzy, a bright violet that darkens when rubbed. Spines up to 5 mm long, crowded, fine, dirty white and becoming brown. **Spores** light brown in deposit, subglobose to elliptical, warted, 4–6 x 3–4 µm.

HABITAT AND ROLE: *Hydnellum suaveolens* is found scattered under spruce wherever the latter is found in the Alaska interior. The coloration of the basiome's interior is particularly striking. When very young, and before the cap is entirely developed, the fungus appears as a dark violet clump of fuzz on the soil.

EDIBILITY, TASTE, AND ODOR: The texture of this fungus is so tough that it is not considered edible.

FAMILY: Bankeraceae (Hydnaceae)

GENUS AND SPECIES: *Sarcodon imbricatus* (L.) P. Karst. (1881) (Fig. 124a–b)
= *Hydnum imbricatum* L. (1753)

COMMON NAME: Scaly tooth

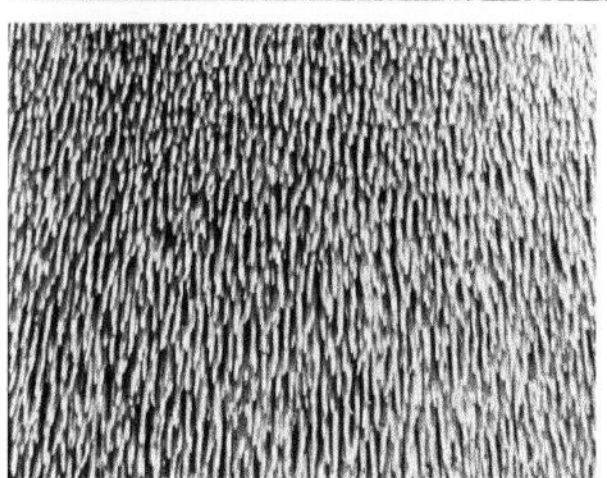

Figure 124a–b *Sarcodon (Hydnum) imbricatus* (L.) P. Karst. (1881), (a) basidiomes (b) hymenial spines

STRIKING FIELD CHARACTERS: Mushroomlike fruiting bodies, light to dark tan with large (and darker) recurred cales all over cap and in concentric rings, spines white, tan, or brown.

MACRO AND MICRO DESCRIPTION: **Cap** of fruiting bodies up to 15 cm across, circular to lobed, convex to flattened, with depressed center or sometimes with a central hole; scaly from the beginning; scales in concentric rings, more or less dark brown on a tan background. Stem 5–12 by 0.7–2 cm, central to somewhat off-center, stocky to slender, equal or somewhat enlarged below, fibrillose, hairless (glabrescent). Spines are up to 9 mm long, extending down stem, crowded, nearly white at first but brown in age, flesh pale brown to tan in the cap, somewhat darker in the stem. **Spores** brown in deposit, subglobose, warted, 6–8 x 5–7 µm.

HABITAT AND ROLE: *Sarcodon imbricatus* occurs in spruce or in mixed forests of the Alaska interior. The species is particularly common in areas near well traveled areas.

EDIBILITY, TASTE, AND ODOR: *Sarcodon imbricatus* is not recommended for eating because of its tough texture. Another look-alike, *S. fennicus* (P. Karst.) P. Karst. (1887), has a stipe base context tissue that is bluish (cut the whole fruitbody in half) and a bitter taste with a tendency to cause gastrointestinal upset.

FAMILY: Hericiaceae

GENUS AND SPECIES: *Hericium coralloides* (Bull.) Letell. (1826) (Fig. 125)
= *Hericium ramosum* (Bull.) Letell. (1826)

Figure 125. *Hericium coralloides* (Bull.) Letell. (1826)

COMMON NAME: Comb tooth

STRIKING FIELD CHARACTERS: Fruiting bodies white to yellowish tan (in age), amorphous to somewhat globose, with pendent spines hanging from branching "limbs," attached to log butts or standing trees.

MACRO AND MICRO DESCRIPTION: Caps of fruiting bodies are clusters (sometimes up to 20–30 cm diameter) of white branches up to 20 cm long. **Stalk** arising from a basal stub. Spines (teeth) are white and hang from the undersides of the branches from base to tips. Texture of both branches and spines is soft at first with spines becoming brittle. The stalk base is inconspicuous and the branches are up to 1 cm diameter, and the teeth are up to 2 cm long. **Spores** white in deposit, subglobose, smooth to minutely roughened, amyloid, 3–5 x 3–4 µm.

HABITAT AND ROLE: *Hericium coralloides* is easily identified and occurs on (or in) the wood of mostly downed hardwood logs, but occasionally on standing trees and stumps, occasionally from wounds of living trees (especially birch).

EDIBILITY, TASTE, AND ODOR: *Hericium coralloides* is an edible fungus that is quite good.

Coraloid (Club) Fungi

The coral fungi that one is likely to encounter in the Alaska interior are white, yellowish olive, dull tan, to bright yellow in color (see Fig. 16e). The fruiting bodies often project upward through the forest floor duff or from rotting wood and take the forms of single "clubs"; a straight to branched, "coral-like" growth form; or a dense cauliflower-like structure resembling the underwater marine corals of the Caribbean or South Pacific. Here their resemblance ends, because these fungi are soft, pliable to brittle-fleshy to the touch. The hymenium, like that of the tooth fungi, consists of a thin layer over the surface of the upright branches. All of the examples considered herein belong to one family, the Clavariaceae.

These fungi could probably be eaten, but one is cautioned against it as some species may cause gastrointestinal upset and/or diarrhea. In general, only the larger corals tend to cause the latter, while the smaller corals are infrequently found, at least in great enough quantity to make a suitable meal.

FAMILY: Auriscalpiaceae (Clavariaceae)

GENUS AND SPECIES: *Clavicorona pyxidata* (Pers.) Doty (1947) (Fig. 126)

COMMON NAME: Crown-tipped coral

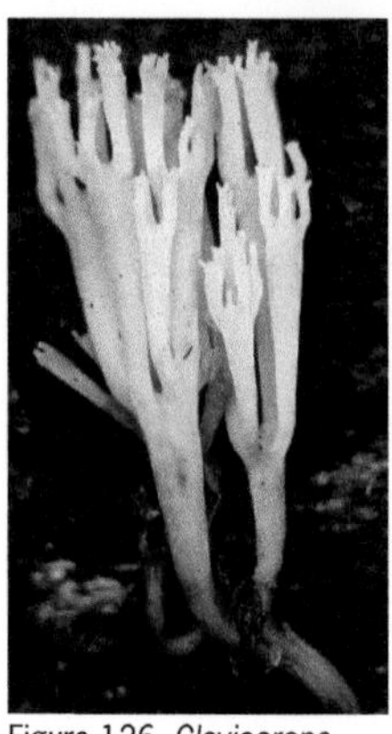

Figure 126. *Clavicorona pyxidata* (Pers.) Doty (1947)

STRIKING FIELD CHARACTERS: Candelabra-like fruiting body with upright branching; dull creamy white with pink to wine tints, darker over the base with basal hairs; on wood, often wood that is buried.

MACRO AND MICRO DESCRIPTION: Coral body 2–4 cm broad, 2–7 cm tall, many branches in groups of 3 or 4 with crownlike tips, dull yellow to cream-colored with pinkish tints above and vinaceous (wine red) tints below, darker over hairy base (use your hand lens). **Spores** white in deposit, elliptical, smooth, amyloid, 3.5–6 x 2–3 µm.

HABITAT AND ROLE: *Clavicorona pyxidata* is a wood decomposer found growing on old, rotting, and often charred balsam poplar logs that have been down for a long time and grown over by a moss carpet that has since been burned off due to old forest fires or ground fires.

EDIBILITY, TASTE, AND ODOR: The taste of this fungus is described as acrid (peppery), but it is reported to be edible.

FAMILY: Gomphaceae (Clavariaceae)

GENUS AND SPECIES: *Clavariadelphus ligula* (Schaeff.) Donk (1933) (Fig. 127)

COMMON NAME: Hairy-based fairy club

Figure 127. *Clavariadelphus ligula* (Schaeff.) Donk (1933)

STRIKING FIELD CHARACTERS: Simple small to medium-sized clubs, yellow, with whitish hairy bases.

MACRO AND MICRO DESCRIPTION: **Fruiting body** of simple clavate clubs, erect and unbranched, but sometimes irregularly forked, terete (rounded cylindrical) to compressed (flattened along their length); dull yellow to yellowish tan or with reddish to vinaceous (wine red) tints; apex bluntly pointed; 3–8 cm by 3–12 mm broad, and mostly smooth to lightly wrinkled. **Spores** white to pale yellow in deposit, elliptic to elongated, smooth, 8–18 x 3–6 µm.

HABITAT AND ROLE: *Clavariadelphus ligula,* when fruiting, forms gregarious troops over the floor in mixed deciduous conifer forests. These and other similar club fungi are thought to be decomposers, but some species are suspected to be mycorrhizal.

EDIBILITY, TASTE, AND ODOR: The taste of this fungus is bland, and the odor mild, but edibility is, once again, best left to the bugs and slugs.

FAMILY: Gomphaceae (Clavariaceae)

GENUS AND SPECIES: *Clavariadelphus pistillaris* (L.) Donk (1933) (Fig. 128)

COMMON NAME: Common club coral

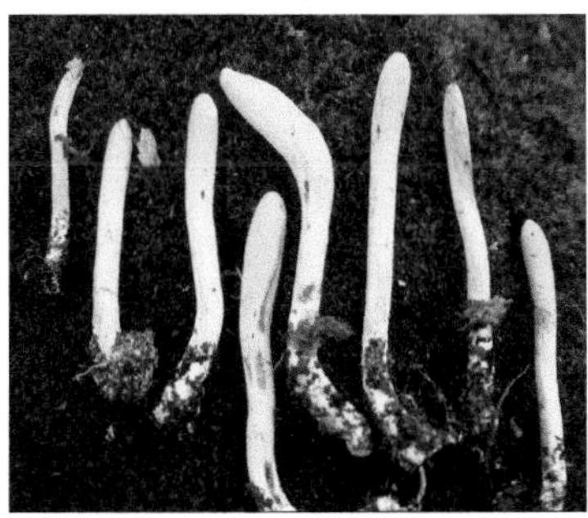

Figure 128. *Clavariadelphus pistillaris* (L.) Donk (1933)

STRIKING FIELD CHARACTERS: Medium to large pestle-shaped, yellow to pinkish tan, smooth to wrinkled clubs. This fungus is macroscopically reminiscent of other species of *Clavariadelphus*; *C. subfastigiatus* is rather similar in appearance, but has smaller spores.

MACRO AND MICRO DESCRIPTION: Fruiting body clavate or club-shaped, and tapering toward base, rounded to somewhat flattened at the apex, 4–10 cm tall and 1–5 cm broad; surface smooth to minutely wrinkled or broadly "grooved"; color white to dull white at base, becoming pinkish white, pinkish tan, yellow-brown to reddish brown over apex; tissue staining buff to pinkish tan. **Spores** white to yellow-tinged in deposit, elliptic, smooth, 9–16 x 5–10 µm.

HABITAT AND ROLE: Solitary to scattered to sometimes gregarious in mixed conifer forests (with aspen and birch), growing on rich humus and duff.

EDIBILITY, TASTE, AND ODOR: *Clavariadelphus pistillaris* is reported as "harmless" (i.e., edible), but this species is hardly worth the effort, as it has a bitter taste and no distinct odor.

FAMILY: Gomphaceae (Clavariaceae)

GENUS AND SPECIES: *Ramaria formosa* (Fr.) Quél. (1888) (Fig. 129)

COMMON NAME: Handsome yellow-tipped, or pinkish coral

Figure 129. *Ramaria formosa* (Pers.) Quél. (1888)

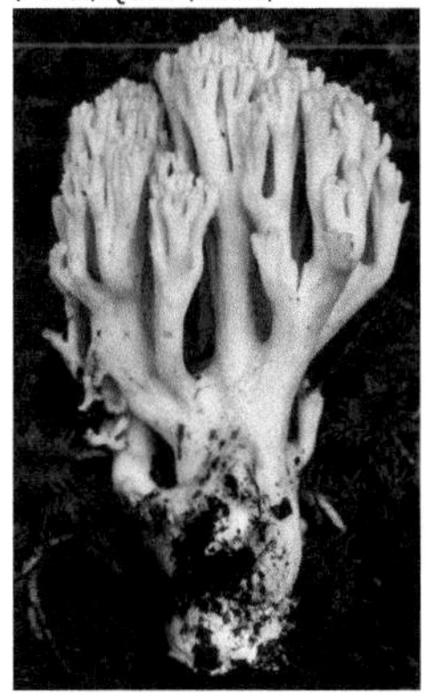

STRIKING FIELD CHARACTERS: Forming dense clusters of upright, crowded branches; 7–15 cm high; dull cream, tan to faintly pink.

MACRO AND MICRO DESCRIPTION: Fruiting body profusely branched (see Fig. 16e), from a large stalk, 7–15 cm high and as wide; hymenium (branches) smooth to gently grooved, erect to flaring (spreading out); dull yellow, cream-colored below to pink or with vinaceous (wine red) tinges above; **stalk** often massive in relation to branches, white fibrous and not gelatinous. **Spores** pale ochraceous in deposit, ellipsoid, roughened, 9–12.5 x 3.5–6 µm.

HABITAT AND ROLE: On rich forest floor humus or duff of mixed white spruce–deciduous forests.

EDIBILITY, TASTE, AND ODOR: The odor of this fungus is mild, and the taste bitter, but it should not be eaten because of its gastrointestinal-upsetting qualities.

FAMILY: Gomphaceae (Clavariaceae)

GENUS AND SPECIES: *Ramaria* aff. *rasilispora* Marr & D.E. Stuntz (1974) (Fig. 130)

Figure 130. *Ramaria* aff. *rasilispora* Marr & D.E. Stuntz (1974)

COMMON NAME: Yellow coral

STRIKING FIELD CHARACTERS: Short thick branches, yellow above and white below.

MACRO AND MICRO DESCRIPTION: **Fruiting body** thickly and abundantly branched; branches bright yellow over upper half, yellow to sometimes white below and not staining; tissue brittle-fibrous. **Spores** pale orange-yellow in deposit, cylindrical, roughened, 8–12 x 3–4.5 μm.

HABITAT AND ROLE: This coral fungus is a member of a group with many possible species. Whether or not any of the others are found in the Alaska interior is questionable, and needs additional work. Mostly solitary, this fungus has been found on the ground in mixed conifer forests.

EDIBILITY, TASTE, AND ODOR: Because we are not sure of the exact identity of this taxon, we encourage caution.

Cantharelloid Fungi

Only two members of the chanterelle family, the Cantharellaceae, are found with any regularity in the Alaska interior, and even then it is rather inconspicuous (see Fig. 16f). One fungus (*Arrhenia lobata*) is most often associated with the various mosses, especially members of the genus *Plagiomnium* or *Rhizomnium*, typically found along edges of shaded streams, rivulets, and water passages. *Mnium* is recognized by its large, rounded leaves. The habitat types in which this moss occurs are very common in the Alaska interior. The other, *Chantharellus infundibuliformes*, is most often associated with deep mosses (*Hylocomnium splendens* and *Pleurozium schreberi*) of dense canopied white and black spruce forests.

FAMILY: Tricholomataceae

Figure 131. *Arrhenia lobata* (Pers.) Kühner & Lamoure ex Redhead (1984)

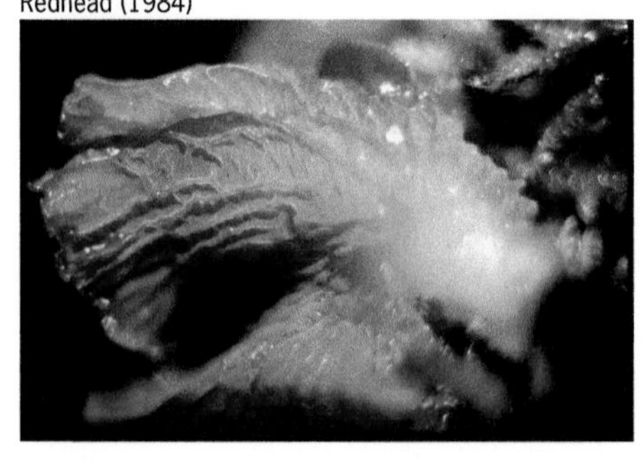

GENUS AND SPECIES: *Arrhenia lobata* (Pers.) Kühner & Lamoure ex Redhead (1984) (Fig. 131)

= *Leptoglossum lobatum* (Pers.) Ricken (1915)

COMMON NAME: Gray-brown chanterelle

STRIKING FIELD CHARACTERS: Cap gray brown; spathulate (spoon-shaped), thinly lobed, and veinlike ridges for "gills"; stalk laterally attached to stream-side mosses.

MACRO AND MICRO DESCRIPTION: **Cap** 0.5–2 cm broad and 0.5–1.5 cm long, smooth, moist but not tacky, gray-brown, subzonate, concentrically lobed to spatula-like; margin often incurved. Ridges rudimentary gills, veinlike, blunt, anastomozing ridges, grayish tan. **Stalk** absent or very rudimentary. **Spores** white in deposit (hyaline), short elliptic (lacrymoid), smooth 7.5–10 x 5–7 µm.

HABITAT AND ROLE: *Arrhenia lobata* is a saprophyte of wet and older moss stems and leaves, occurring along rivulets, brooks, and moss-lined streams. This fungus is inconspicuous, small, fragile, and not easily seen. As such, it is a "hands and knees" fungus, most often found as single fruiting bodies. *Arrhenia lobata* may be gregarious, in small clumps, but the species is rarely abundant.

EDIBILITY, TASTE, AND ODOR: The odor of *Arrhenia lobata* is fishy or like that of the moss to which it is attached. Its taste is watery-bland, and being so small, inconspicuous, and few in number, is hardly something one would go looking for to gather for the table.

Thelephoroid (Vase) Fungi

Members of the genus *Thelephora* (see Fig. 16g), which is placed into its own family, the Thelephoraceae, are often very common as dark purple brown "vases" 6–25 mm high, typically almost equally wide, and found on disturbed, thin, or sparsely moss-carpeted soils in close association with pioneer plants such as willows and alders along riparian stream courses. The hymenium covers the outside of the vase above whatever stalk may be present. Most often the hymenophore is smooth to wrinkled, green-brown to red-brown, or sometimes deep chocolate brown to purplish-brown.

These fungi are quite small and even though they often carpet the ground in places, we usually never think of them as a potential food source.

FAMILY: Thelephoraceae

GENUS AND SPECIES: *Thelephora terrestris* Ehrh. (1786) (Fig. 132a–b)

COMMON NAME: Common fiber vase

STRIKING FIELD CHARACTERS: Small, tough, brown, stalked, and vase-like mushrooms with smooth, warted or rippling, gray underside.

Figure 132a–b. *Thelephora terrestris* Ehrh. (1786), (a) fan (b) cup

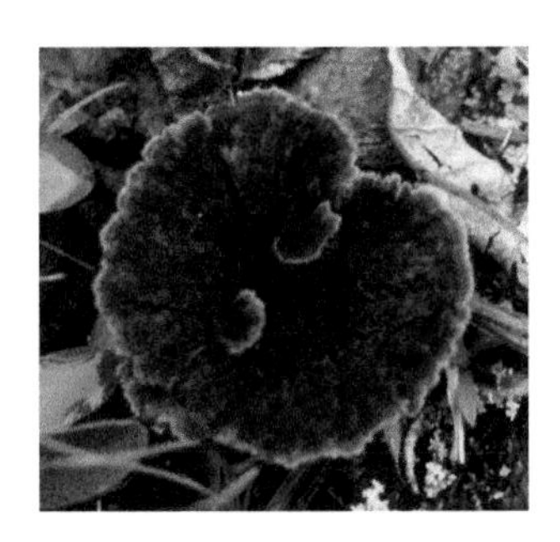

MACRO AND MICRO DESCRIPTION: **Fruiting bodies** with a cap up to 6 cm in diameter, circular to semicircular, **stalk** 1–3 cm long, cap hairy to scaly, reddish brown to dark brown; margin rough and toothy, lower surface irregular, smooth to warted, pinkish gray to reddish brown. **Spores** purple-brown in deposit, ellipsoid-angular, warted, 8–12 x 6–9 μm.

HABITAT AND ROLE: This species is a mycorrhizal associate of willows and spruce in the Alaska interior and is not uncommon on exposed soil beneath other conifers.

EDIBILITY, TASTE, AND ODOR: *Thelephora terrestris* is not considered edible.

Fungal Groups: The Gasteromycetes

(Having No Organized Hymnenium)

Gasteroid (Stomach) Fungi: Puffballs

Most often referred to as Gasteromycetes, the stomach fungi abound in the Alaska interior. Puffballs (see Fig. 16h), the most commonly encountered members of the group, occur as single fruiting bodies or are gregarious to forming tufted masses of fruitings on moss-covered soil, on woody debris or rotting logs, along the forest floor, and amongst disturbed camp sites. Only three genera (*Bovista*, *Calvatia*, and *Lycoperdon*), all members of one family, the Lycoperdaceae, are considered here. There are several other forms, but these are small and rather inconspicuous and/or uncommon.

Perhaps the easiest rule of thumb for determining edibility for Interior fungi pertains to this group, the puffballs. If preparing puffballs for the table, cut them in half with a sharp knife to make sure the puffball you have collected is actually just that and not a very young, developing, and unexpanded agaric. If white throughout, then the puffball is edible. If the puffball interior (**gleba**) is changing color (from white to yellow, greenish yellow, olive, rusty, brown, or chocolate), then you should refrain from eating it. At full maturity, the interior will be dry, powdery, and impossible to eat without choking. The outer wall (**peridium**) may have a hole-like opening (**ostiole**) on top, be deeply cracked, or simply badly weathered. This is necessary for spore discharge, as the spores of these fungi—unlike those of agarics, boletes, polypores, tooth fungi, coral fungi, chanterelles, and many cup fungi—are not forcibly discharged. Therefore, spore discharge in the puffball is affected by animals stepping on fruiting bodies, small animals carrying away spores stuck to their fur, by the impact of raindrops, or by simple erosion of parts for eventual wind dissemination of spores.

FAMILY: Lycoperdaceae

GENUS AND SPECIES: *Calvatia booniana* A.H. Sm. (1964) (Fig. 133)

COMMON NAME: Giant alpine puffball, Western giant puffball

STRIKING FIELD CHARACTERS: Large egg-shaped, white to yellowish brown, sessile fruiting body; outer surface "warts" patterned, flattened with tan to yellow-brown centers and convex sterile base rising up into a "ball" from an attachment point; on meadow tundra.

Figure 133. *Calvatia booniana* A.H. Sm. (1964), (a) external (b) internal

MACRO AND MICRO DESCRIPTION: **Fruiting body** large, 8–10 cm high, 15–18 cm long, "flattened" egg-shaped, upper half of surface covered with low, flattened scale-like "warts," wrinkled to smooth below to point of ground attachment. Inner **gleba** white and firm in youth becoming soft-spongy with age; becoming olive yellow, olive-brown, and finally dark brown at maturity. **Sterile base** remains white, broadly lenticular (convex lens–shaped) one-third to one-half the total height. **Spores** olive-brown, round to broadly elliptical, smooth to minutely ornamented, 4–6.5 x 3.5–5.5 µm.

HABITAT AND ROLE: *Calbovista subsculpta* is a fungus that has been found twice at Sable Pass on a hillside ca. 100 m south of the road, where it occurred in a mesic alpine tundra meadow herb field. This fungus is a decomposer of herbaceous detritus (dead plant remains), and is the largest puffball we've ever found in Alaska.

EDIBILITY, TASTE, AND ODOR: Both taste and odor of *c.b.* are mild. It is edible and good if fresh and solid white. Once it begins turning greenish yellow to olive brown it also becomes "water logged." At maturity you would never want to eat it, lest you choke to death trying! Another close look-alike, *Calboviata subsculpta*, which can also be found in high-altitude alpine meadows, is smaller, and the warts are more "raised" if not pyramidal and with a brown gleba.

FAMILY: Agaricaceae (Lycoperdaceae)

GENUS & SPECIES: *Bovista plumbea* Pers. (1795)

COMMON NAME: Tumbling puffball

STRIKING FIELD CHARACTERS: The fruiting body is globose, subglobose to flattened egg-shape, smooth to patchy areolate, white in youth to dirty-gray to light tan, then becoming brownish-gray (steely) and "papery" thin at maturity, base absent, but often with a persistent rhizomorphic mycelial and centrally attached substrate penetrating cord. The white layer (outer peridium) soon cracks, exfoliates, and peels off leaving the thin shiny spore sac (inner peridium).

Figure 134. *Bovista plumbea* Pers. (1795)

MACRO & MICRO DESCRIPTION: **Fruiting body** relatively small, 2–4 cm; peridium smooth to having flattened "warts" (areolate), smooth, thin and silver-brown to brownish-gray beneath and in age. **Gleba** (capillitial threads) white to yellow-green and finally rusty-brown. **Sterile base** absent. **Spores** olive brown to deep chocolate, ovoid, broad elliptic to round, minutely spiny to almost smooth, 5–7 x 4.5–6 μm, with long attached pedicels, 5–12 μm long.

HABITAT AND ROLE: This fungus fruits above ground on grassy knolls, lawns and open hay fields here in the Interior. As a decomposer, its most probable role is that of breaking down dead grass leaves and stems. It fruits in late June to mid July.

EDIBILITY, TASTE AND ODOR: As long as the gleba of this puffball is pure white and not discoloring or soggy-squishy, then it is edible. Be sure to cut this fungus open, making sure that it is indeed a puffball, and not an immature agaric before attempting to prepare it for your table.

FAMILY: Lycoperdaceae

GENUS AND SPECIES: *Lycoperdon perlatum* Pers. (1796) (Fig. 135)
= *Lycoperdon gemmatum* Batsch (1783)

COMMON NAME: Gemmed or common puffball

STRIKING FIELD CHARACTERS: Stalked puffball with a sterile base, white to buff to pinkish buff, with deciduous cone-shaped spines that leave scars, terrestrial.

MACRO AND MICRO DESCRIPTION: **Fruiting body** top-shaped with a stemlike base, 1–4 cm broad and 2–5 (–6) cm high; outermost surface with slender, "spatially arranged" short cone-shaped spines and granules that leave scars or pockmarks when rubbed off. Peridium white to dull white, with a pinkish to brownish cast; spines white to gray to brown; inner layer of peridium darkening in age to tan, yellowish brown, or dark brown when old; peridium ruptures by means of an ostiole (pore) at the top center of the fruiting body. **Sterile base** well developed, minutely chambered below the glebal mass; white and spongy when young, then yellowing and finally turning brown. **Gleba** white at first, then yellow, olive, brown, and powdery dry at maturity. **Spores** dark olive-brown to chocolate, round, smooth to spiny, 3.5 x 4.5 μm.

Figure 135. *Lycoperdon perlatum* Pers. (1796)

HABITAT AND ROLE: This terrestrial puffball may be solitary, gregarious, or (most often) clustered in white spruce forests, along paths, trails, and disturbed roadbeds. It fruits in July and August and is most likely a humus decomposer (saprophyte).

EDIBILITY, TASTE, AND ODOR: This is another fine edible puffball that needs only to be white throughout on the inside and devoid of creepy crawlers. Taste is mild to bland and may even be bitter, while the odor is pleasant.

Gasteroid (Stomach) Fungi: Bird's Nest

Bird's nest fungi (see Fig. 16i) are gasteromycetes and thus close relatives of the puffballs. In fact, they are "puffballs" with a slightly different mechanism for spore discharge. The puffballs contain a cinnamon brown, rusty brown, to purple-brown dry powdery spore mass that "puffs" out through a hole (**ostiole**), through cracks, or directly from weathered fruiting bodies being eaten by insects and other small animals, or by being stepped on by big animals such as you, me, bears, moose, etc. Sometimes the name "booters" is applied to puffballs, especially those larger forms, since they "puff" when booted. However, bird's nest fungi rely mainly on big raindrops to disperse their small, egglike sacs of spores (**peridioles**).

Bird's nest fungi derive their name from the appearance of the fruiting bodies when the lidlike covering (**epiphragm**) over the natural opening weathers away to expose those egglike flattened sacs of spores. The resulting cup-shaped basidiocarps employ raindrops to "splash" the peridioles out of the cup. Peridioles catch onto or adhere to surrounding vegetation, where they hang or stick, dry out, and ultimately crack open to release the spores, which are carried away by the wind. This mechanism is very effective, so if you find a few birds' nest fungi on sticks or woodchips, look around closely and you are likely to see a lot more. The peridioles can adhere to green plants and be eaten by grazing animals—probably common in *Crucibulum vulgare* and some of the *Cyathus* spp.

Bird's nest fungi are simply too small to allow one to collect enough to make any kind of a satisfying meal.

FAMILY: Nidulariaceae

GENUS AND SPECIES: *Nidularia* aff. *farcta* (Roth) Fr. (1823) (Fig. 136)

COMMON NAME: Bird's nest fungus

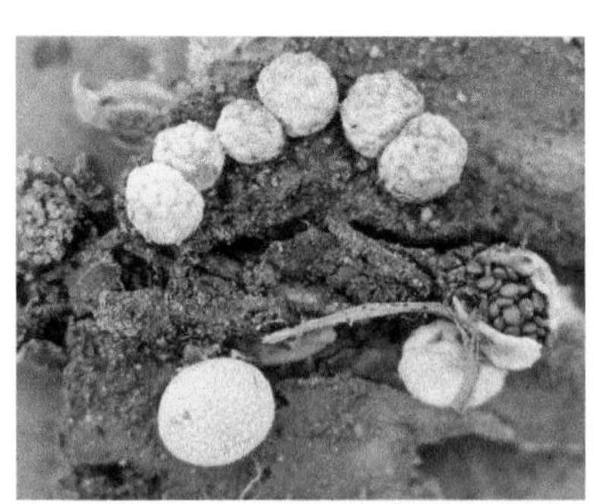

Figure 136. *Nidularia* aff. *farcta* (Roth) Fr. (1823)

STRIKING FIELD CHARACTERS: Tiny cuplike "nest" containing reddish brown, lenticular to flattened **peridioles** ("eggs") that wrinkle when dried. The opening of the "nest" is covered in youth by a feltlike layer that cracks open indiscriminately; on rotting woody debris.

MACRO AND MICRO DESCRIPTION: **Fruiting bodies** small; cushion-, cup-, or flowerpot-shaped; 3–5 mm high to 6–10 mm broad at the top; no "true" lid, only a white to buff, scurfy to hairy exterior covering that splits, crumbles, or opens irregularly; nest wall (peridium) soon disintegrating; white to yellowish tan. **Peridioles** several, 1–2 mm broad, flattened, reddish brown, with a mucilaginous jellylike matrix at the base, drying to "release" the spores. **Spores** hyaline 6–10 x 4–7 µm.

HABITAT AND ROLE: *Nidularia* aff. *farcta* most probably a decomposer of woody substrates and seems more likely to be a cellulose/hemicellulose rather than a lignin decomposer, as the wood substrate (white spruce) becomes brown and crumbly (lignin remains) rather than white and punky (cellulose remains). This fungus will be found in moist dense forests under brush or on bits and pieces of wood debris.

EDIBILITY, TASTE, AND ODOR: Do not even try them. Edibility is purely academic, because of the small size. They are not poisonous, but why bother!

Gasteroid (Stomach) Fungi: Earthstars

The earthstars (see Fig. 16j) are an unusual lot in that they are Gasteromycete (stomach) fungi with an outer peridium (wall) that more often than not splits in a stellate fashion with numerous rays. These outer wall rays repeal back and away from the inner puffball-like sac and form what appear to be star-shaped (stellate) arms that may even "lift" the central portion off the ground, if it is not already on a pedestal. These curious Gasteromycetes, like their close relatives the puffballs, depend on animal consumption, stomping, or raindrops banging on their peridial walls for nonforcible ejection of spores. There are many species of earthstars, but interior Alaska seems to have but four species thus far collected. All four belong to the genus *Geastrum* in *G. coronatum* Pers. (1801), *G. mammosum* Chevall. (1826), *G. xerophilus* Long (1942), and an unknown, but much larger, species.

FAMILY: Geastraceae

GENUS AND SPECIES: *Geastrum mammosum* De Toni (Fig. 137)

COMMON NAME: Earthstar

STRIKING FIELD CHARACTERS: A very small, puffball-like fungus with an outer "skin" that peels back and away from an inner bulb and splits into several star-shaped rays that "lift" the central operculate bulb in dry weather, but flatten in moist weather.

Figure 137. *Geastrum mammosum* De Toni

MACRO AND MICRO DESCRIPTION: Bulb-shaped inner peridium 5–15 mm broad, gray-brown at first to almost blackish, but drying lighter and fading to almost grayish white; peristome prominent with distinct encircling depression. Outer peridium finely felted, thin, soft when moist, hygroscopic, rays 8–12, unequal widths, darker than inner peridium. **Spores** spherical, finely warted, 3–4 µm.

HABITAT AND ROLE: These unique, very small earthstars are associated with grasses on precipitous south- to southwest-facing slopes and more than likely are saprophytes of the dead grass.

EDIBILITY, TASTE, AND ODOR: They are too small to even consider eating. Anyway, you might find the outer peridium rather tough! This one is best left to the rodents.

Fungal Groups: Conifer Rust, Jelly, Cup, and Earth Tongue Fungi

Conifer Rust Fungi

The rusts (see Fig. 16k) belong to a group of Basidiomycetes very different from traditional mushrooms, and most likely much more primitive. These rather distinctive fungi can be found everywhere in forests within the Alaska interior. Many species are represented, but two will most probably stand out to the casual observer. Rust fungi often have very complex life cycles involving plants other than the primary or economically important spruce trees upon which they are most conspicuous. Look carefully at leaves of shrubby understory plants like Labrador tea and bearberry for black, blue, red, cinnamon, orange, or yellow spots. You are probably looking at another stage of the rust on its alternate host if you see these spots. Some of these spots may also be caused by viruses or insects, which are not of interest to us here.

Two rusts will be seen abundantly in the Alaska interior. These are the spruce needle and spruce broom rusts, on white and black spruce.

FAMILY: Coleosporiaceae (Pucciniaceae)

GENUS AND SPECIES: *Chrysomyxa ledicola* Lagerh. (1893) (Fig. 138)

Figure 138. *Chrysomyxa ledicola* Lagerh. (1893)

COMMON NAME: Spruce needle rust

STRIKING FIELD CHARACTERS: Premature defoliation of yellow-green, yellow, to yellow-brown needles, especially at the tips of branches having white pustules on the undersides of the needles.

MACRO AND MICRO DESCRIPTION: **Spores** (aecial) warted, 22–34 x 27–46 µm, on Labrador tea upper leaf surface; (uridiniospores) warted, 20–33 x 24–45 µm, on previous year's spruce; (teliospores) 10–14 x 13–18 µm, on spruce.

HABITAT AND ROLE: *Chrysomyxa ledicola* is a common foliage rust of white and black spruce in the Alaska interior (and also Sitka spruce in southeast Alaska), where the ericaceous ground cover, Labrador tea, occurs in association with spruce. This rust causes yellow discoloration of branch-tip needles and premature needle loss, but its presence does not generally result in the death of the tree unless the location is totally infected or heavily infected for multiple years, but does cause premature defoliation and, rarely, cone rust.

Figure 139a–b. *Chrysomyxa arctostaphyli* Dietel (1894), (a) bole (b) broom

FAMILY: Pucciniaceae

GENUS AND SPECIES: *Chrysomyxa arctostaphyli* Dietel (1894) (Fig. 139a–b)

COMMON NAME: Spruce broom rust

STRIKING FIELD CHARACTERS: Look to the tops of white or black spruce for branch proliferations into large ball-like orange witches'-brooms, from which needles are shed after turning yellow, then orange, and finally rusty orange. It causes purple-brown leaf spot on its alternate host, kinnikinnick.

MACRO AND MICRO DESCRIPTION: **Spores** (teliospores) 13–18 x 23–64 µm, on spruce needles; (basidiospores) 8.5–9.5 x 7.5–8 µm.

HABITAT AND ROLE: *Chrysomyxa arctostaphyli* infects an ericaceous ground-cover plant, bearberry (*Arctostaphylos uva-ursi*), causing a purple-brown leaf spot on this "alternate" host. Spruce is the primary host, the one most affected. These brooms provide nest-building sites and ultimately fungal basidiocarp-caching sites for the northern red and flying squirrels. Often, the fungi they pick from the forest floor are epigeous (aboveground) or hypogeous (belowground), or are ectomycorrhizal species of agaric and false truffle fungi. Once dried in the limbs of the tree, fruiting bodies are stored for winter food in the old nest sites. *C. arctostaphyli* may cause growth stunting (in height and diameter), bole (trunk) deformation, knots, and even tree mortality.

Jelly Fungi

The jelly fungi (see Fig. 161) often abound in nature, but one really has to look carefully to find them in the Alaska interior. They all grow on wood; some live on hardwoods such as alder and birch, while others grow on dead wood branches of spruce and willow, particularly at slightly higher elevations. They are an intriguing lot, and their fruiting bodies are most conspicuous right after a prolonged rainy, foggy, or wet period. Often found to exhibit bright colors such as black, red, orange, and yellow (the most commonly encountered), jelly fungi have very little "shape" and look like amorphous globs of jelly on their selected substrates. Surprisingly, however, they do have "integrity" in that they do not smear and stick to you like, for example, a fresh plasmodium of a slime mold. They actually feel like your earlobe when fresh. When dry, they often turn black, shrivel to practically nothing, and appear as yellow to brown or blackened dried gluelike blobs or crusts on woody substrates.

These fungi are Basidiomycetes just like the mushrooms and toadstools, but their microscopic architecture is different; hence, the taxonomic position and placement into their own separate group of fungi. There are several species in the Alaska interior, but we have selected perhaps the most conspicuous form present in the Alaska interior to serve as an example of the group. Their role is no less important than some of the other Basidiomycetes, as they are saprophytic fungi (or mycoparasites) that decompose dead wood (sticks, branches, and fallen logs) and, with time, clean up the woody debris on the forest floor. *Tremella* spp. and other members of the Tremellaceae are probably all mycoparasites; they parasitize homobasidiomycetes of various types, lichen fungi, and others. They, like other Hymenomycetes, have an organized hymenium.

FAMILY: Tremellaceae

GENUS AND SPECIES: *Tremella mesenterica* Retz. (1769) (Fig. 140)

COMMON NAME: Witch's butter

STRIKING FIELD CHARACTERS: Yellow jellylike (when fresh) blobs on alder.

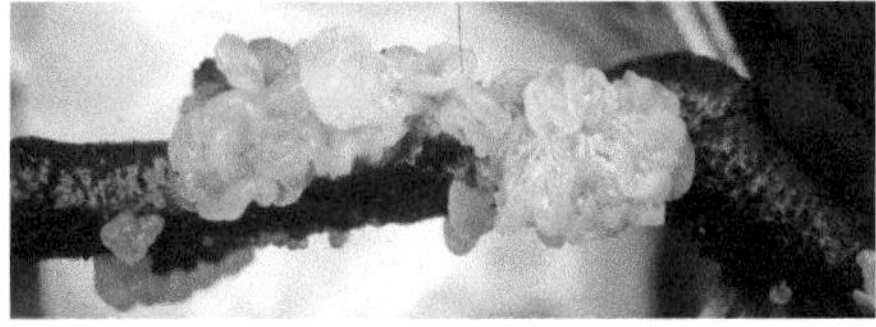

Figure 140. *Tremella mesenterica* Retz. (1769)

MACRO AND MICRO DESCRIPTION: **Fruiting bodies** tough, jellylike, irregularly shaped globs or lobes 2.5–10 cm wide and 3–4 cm high; translucent pale yellow to yellow-orange, changing to orange-red on drying. **Spores** pallid to yellow, broadly elliptical, globose to subglobose, smooth, 7–18 x 6–14 µm.

HABITAT AND ROLE: *Tremella mesenterica* is found rather commonly on small dead hardwood branches, especially those of alder, and are usually associated with *Peniophora* spp. and other early colonizers of wood.

EDIBILITY, TASTE, AND ODOR: *Tremella mesenterica* is sometimes said to be edible as an additive to soups, but it is not commonly eaten. *T. mesenterica* and other species with large basidiomes are eaten by many East Asians. Two species are cultivated in large quantities: *T. fuciformis* is grown in Japan, China, and Korea, primarily for use as food. Some of the other Tremellas are used primarily as medicinals.

Cup Fungi

The cup fungi, unlike all of the other groups considered thus far, do not produce their spores on the outside of basidia (see Fig. 16m). Instead, they produce their spores inside elongated, hollow, sausage-shaped cells (sacs) called **asci**, hence the name ascomycete. These fungi are often small, fairly inconspicuous, or hidden from view, even though some are brightly colored. Many do forcibly discharge their spores. When you find one, blow (breathe) very gently across the cup and watch for a light-colored "smoke" that emanates from the colored hymenium that lines the cup. Unlike the Thelophores, which resemble some cup fungi, the spore-producing surface (hymenium) is found **inside** the vase or cup. Of course, the spore-producing cells (asci) differ significantly too. Five genera in as many families (Dermatiaceae, Pyronemataceae, Helvellaceae, Hypocreaceae, and Pezizaceae) commonly found in the Alaska interior are considered herein. There are, however, a number of other genera and many more species known to occur throughout the interior of Alaska.

FAMILY: Pyronemataceae

GENUS AND SPECIES: *Geopyxis carbonaria* (Alb. & Schwein.) Sacc. (1889) (Fig. 141)

COMMON NAME: Pixie cup

STRIKING FIELD CHARACTERS: Brown, deep goblet cups with toothed yellow (cream) margin and slender dull white to buff-colored stalk.

MACRO AND MICRO DESCRIPTION: **Cup** 0.5–1.5 cm broad, often as deep as wide; outer surface tan to light brown, becoming darker toward margin that is dissected, cream to yellow-brown and scaly-toothed; inner surface an orange-brown (darker). **Stalk** 0.5–1.5 cm long and 2–3 mm broad, dull white, smooth to wrinkled. **Spores** narrowly elliptical, 12–18 x 6–9 µm.

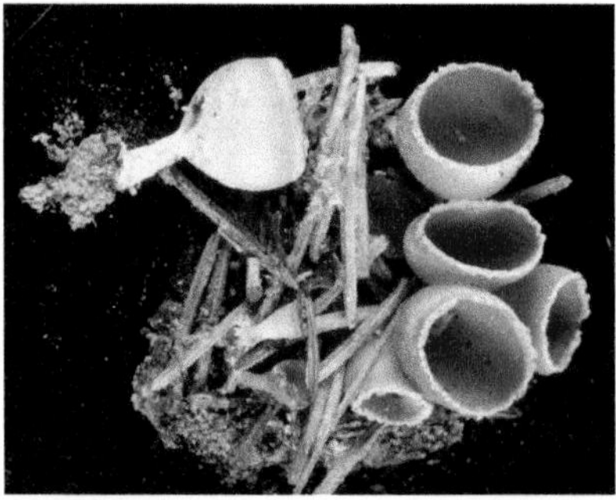

Figure 141. *Geopyxis carbonaria* (Alb. & Schwein.) Sacc. (1889)

HABITAT AND ROLE: After a forest fire, this gregarious cup fungus pops up everywhere on charred white and black spruce needles. This species is close to others in its genus (e.g., *Geopyxis vulcanalis*) and several species in other genera (*Tarzetta* and *Pustularia*) that may ultimately be found in the Alaska interior but so far have eluded us.

EDIBILITY, TASTE, AND ODOR: The edibility of *Geopyxis carbonaria* is not reported one way or the other. Besides, they are so small a mouthful would take hours of collecting.

FAMILY: Helvellaceae

GENUS AND SPECIES: *Helvella corium* (O. Weberb.) Massee (1895) (Fig. 142)

Figure 142. *Helvella corium* (O. Weberb.) Massee (1895)

COMMON NAME: Black scurfy fairy cup

STRIKING FIELD CHARACTERS: Cup small to medium, black, and scurfy (minutely scaly), often with a white scurfy margin, smooth black bowl. Stalk gray to dull white, smooth to ribbed.

MACRO AND MICRO DESCRIPTION: **Cup** 0.5–5 cm broad and often compressed laterally, black scurfy outer surface and smooth black inner (bowl) surface; margin often chalky white. **Stalk** 1–3 cm long and 2–8 mm broad, gray above, dull white at base, and sometimes minutely furrowed. **Ascospores** obtuse ended, 16.5–19 x 9–10.5 µm.

HABITAT AND ROLE: This fungus is one of the few cup fungi that form ectomycorrhizal (fungus-root) associations with dwarf alpine/arctic willows.

EDIBILITY, TASTE, AND ODOR: This fungus has a mild odor and taste, but we know nothing of its specific edibility. It belongs to a group, *Helvella*, with many examples that do contain gastrointestinally upsetting compounds. Do not try it. It's not worth the gamble.

FAMILY: Rhytismataceae

GENUS AND SPECIES: *Rhytisma salicinum* var. *arcticum* P. Karst. (1871) (Fig. 143a–b)

COMMON NAME: Leaf tar spots

STRIKING FIELD CHARACTERS: Black "tar" spots on upper and/or undersides of arctic willow and dwarf birch leaves.

MACRO AND MICRO DESCRIPTION: **Fruiting bodies** take the form of a small, black, erumpent, convex disc, 1–3 mm diameter and "raised" on leaves of dwarf birch. **Ascospores** hyaline, fasciculate, filiform, 60–80 x 1.5–2.5 µm.

Figure 143a–b. *Rhytisma salicinum var. arcticum* P. Karst. (1871)

HABITAT AND ROLE: This fungus is a parasite of circumpolar dwarf birch and is quite common during some years. It typically fruits in late July and early August, just before freezing temperatures result in plant leaf drop.

EDIBILITY, TASTE, AND ODOR: This is not a fungus you would want to eat. Bitter leaf tissues would also overpower whatever flavor might be there.

FAMILY: Hypocreaceae

GENUS AND SPECIES: *Hypomyces chrysospermus* Tul. & C. Tul. (1860) (Fig. 144)

COMMON NAME: Bolete eater

STRIKING FIELD CHARACTERS: First white, then lemon to egg-yolk yellow, and finally dirty yellow to orange-brown; causes host bolete (*Xerocomus* in particular) to become soft and spongy. Mycelial stage is quite common with its large globose, yellow, and warted chlamydospores. In this asexual stage it is called *Sepedonium chrysospermum*.

Figure 144. *Hypomyces chrysospermus* Tul. & C. Tul. (1860)

MACRO AND MICRO DESCRIPTION: **Fruiting bodies** small, not always evident; the vegetative mycelium forms a yellow covering over old fruiting bodies of other fungi. This rather distinctively colored *Hypomyces* is an ascomycete with asci and ascospores produced in flask-shaped perithecia. However, a bright yellow (*Sepedonium*) stage that covers the bolete's pore surface first produces chlamydospores (thick-walled and asexual [conidia] spores). Later, the final orange-brown stage, rarely encountered, produces the sexual ascospores in flask-shaped perithecia. **Ascospores** if white en masse, elliptic, smooth, 10–30 x 5–12 μm; if yellow en masse, globose, thick-walled, warted, 10–25 μm.

HABITAT AND ROLE: A parasite of boletes, *Hypomyces chrysospermus* is quick to overgrow the pore surfaces of maturing boletes.

EDIBILITY, TASTE, AND ODOR: This is reported to be nonedible and may, in fact, be poisonous because of its association with poisonous bolete species or bacterial decay of the bolete tissues themselves. These bacteria may actually be the culprits in that they produce metabolites, which may cause "poisoning."

Figure 145. *Hypomyces tulasneanus* Plowr. (1882)

FAMILY: Hypocreaceae

GENUS AND SPECIES: *Hypomyces tulasneanus* Plowr. (1882) (Fig. 145)

= *Hypomyces luteovirens* (Fr.) Tul. & C. Tul. (1860)

COMMON NAME: Green lobster

STRIKING FIELD CHARACTERS: Yellow-green to green and/or bluish green, on fruitbodies of the Russulaceae (*Russula* and *Lactarius*).

MACRO AND MICRO DESCRIPTION: **Ascospores** fusiform to naviculate, one-celled, nearly smooth to prominently verrucose and apiculate, 32–35 x 4.5–5.5 µm.

HABITAT AND ROLE: Parasitizes agaric fungi in the Russulaceae, first in an all-white mycelial state, but then turning yellow-green dark green, to bluish green at maturity.

EDIBILITY, TASTE, AND ODOR: Some claim the *Hypomyces*-infected basidiomycetes are a real delicacy. For us, we cannot get by the "rotting" factor and find the idea to be absolutely repugnant.

FAMILY: Discinaceae Benedix (Helvellaceae)

GENUS AND SPECIES: *Gyromitra infula* (Schaeff.) Quél. (1886) (Fig. 146)

COMMON NAME: False morel

STRIKING FIELD CHARACTERS: Tan to brown, smooth to wrinkled cap that does not attach to the white stalk.

MACRO AND MICRO DESCRIPTION: **Cap** yellow-brown, reddish brown to dark brown (center top), smooth to folded fertile surface (hymenium), 1.5–4 cm broad and high, lobed, saddlelike or hoodlike to mostly irregularly lobed (not tightly wrinkled), with incurved margin, "attached" in a few places to stalk. **Stalk** hollow, somewhat irregular to flattened, but generally larger toward base, 1–4 cm long and 0.5–1.5 cm broad. **Ascospores** biseriate, narrowly elliptic, smooth, 19–23 x 7–8 µm.

Figure 146. *Gyromitra infula* (Schaeff.) Quél. (1886)

HABITAT AND ROLE: Solitary to gregarious on moss-covered soil of previously burned spruce needle duff; this fungus is a saprophyte, and usually occurs on wood, but ours is on burned peat and conifer duff substrates that are also high in cellulose content.

EDIBILITY, TASTE, AND ODOR: The *Gyromitra infula* "group" is reported to be poisonous, as the fruiting bodies contain the carcinogen monomethylhydrazine (MMH) and other volatile and water-soluble substances that may cause serious health problems.

Earth Tongue Fungi

The earth tongues (see Fig. 16n) are relatively small Ascomycetes in which the spore-producing surface is similar to the basidiomycetous coral fungi, occurring around the outside of the often-flattened, spade-like fruiting bodies that typically project out of mossy substrates here in the Alaska interior. Unlike the cup fungi, their microscopic asci (sacs) have no "lid-covered" opening. Instead, there exists a pore through which spores are forcibly ejected. The earth tongues common to the Alaska interior are stalked, yellowish tan to black, and generally conspicuous when fruiting against a bright green moss background. Members of the group are saprophytes on dead moss parts. The more colorful earth tongues found in Denali National Park belong to the genera *Spathularia* and *Mitrula*, whereas the darker (black) forms belong to genera of *Geoglossum* and *Trichoglossum*.

FAMILY: Sclerotiniaceae (Geoglossaceae)

GENUS AND SPECIES: *Mitrula borealis* Redhead (1977) (Fig. 147)

COMMON NAME: Miniature earth tongue

STRIKING FIELD CHARACTERS: Very small, golden-yellow, smooth to wrinkled, and irregularly flattened pear-shaped head with a white stalk.

Figure 147. *Mitrula borealis* Redhead (1977)

MACRO AND MICRO DESCRIPTION: **Caps** (ascocarp) gregarious or in tight troops, fleshy, 1–4 mm high and 1–3 mm broad, ovoid to flattened, pear-shaped, smooth to having small ridges (rugulose), golden yellow (copper yellow), and dry. **Stalk** terete but flexuous, white, almost glistening, 1–2 mm broad and 10–30 mm long. **Ascospores** hyaline, elliptical to ovoid or broadly cylindrical two-celled, gelatinous sheathed, 10.5–18 x (2.5–) 3–4 (–5) µm.

HABITAT AND ROLE: Generally in wet places on mosses or liverworts along sunken roadbeds or creeks. *Bryoglossum gracile* (Karst) Redhead is a very close relative of *Mitrula borealis* on the basis of microscopic characteristics (including smaller spores).

EDIBILITY, TASTE, AND ODOR: *Mitrula borealis* is not recommended, as it is simply too small to make a meal. Taste is bland (mild) or tasteless and odor is fishy (essentially that of the moss substrate on which it is found).

FAMILY: Cudoniaceae (Geoglossaceae)

GENUS AND SPECIES: *Cudonia circinans* (Pers.) Fr. (1849) (Fig. 148)
= *Leotia circinans* Pers. (1797)

COMMON NAME: Common cudonia

STRIKING FIELD CHARACTERS: Rounded; creamy pink, pinkish tan, to rosy tan; often with a "split" margin; cap folded down over an irregularly flattened, fluted, or ridged darker rosy tan to rosy brown stalk.

Figure 148. *Cudonia circinans* (Pers.) Fr. (1849)

MACRO AND MICRO DESCRIPTION: **Cap** 0.5–2 cm broad, rounded, sometimes splitting over the often-incurving margin and indented on top, smooth to wrinkled, pinkish tan. **Stalk** 3–8 cm long and 3–10 mm broad, darker wine brown, smooth to scruffy, irregularly flattened, fluted, or ridged. **Ascospores** hyaline, needlelike, septate, smooth, 30–45 x 2 µm.

HABITAT AND ROLE: *Cudonia circinans* is a rather small to medium-sized fungus that usually occurs in clusters in the Interior. These clusters may be scattered around, so if you find one, take a good look around at the deep moss-covered humus (soil or often rotting wood) for others.

EDIBILITY, TASTE, AND ODOR: *Cudonia circinans* is reported to be poisonous as it contains monomethylhydrazine toxins. Leave this one for the slugs.

FAMILY: Cudoniaceae (Geoglossaceae)

GENUS AND SPECIES: *Spathularia flavida* Pers. (1794) (Fig. 149)

COMMON NAME: Fairy fan

STRIKING FIELD CHARACTERS: Flattened, spathulate, smooth to wrinkled, light yellow fan-shaped head straddles a darker swollen stalk.

Figure 149. *Spathularia flavida* Pers. (1794)

MACRO AND MICRO DESCRIPTION: **Cap** compressed to flattened, spathulate or fanlike, smooth to wrinkled (sometimes contorted), pale yellow, yellow-tan to brown. **Stalk** is often darker-colored and smooth. **Ascospores** yellow-brown in deposit, (hyaline sublenta) needlelike, cylindric-clavate, smooth, one to many septate, 30–75 (–95) x 1.5–3 µm.

HABITAT AND ROLE: Widely to closely scattered on mosses under white or black spruce.

EDIBILITY, TASTE, AND ODOR: We are unaware of any reports relating to the edibility of the fungus.

Lichenicolous Fungi: Parasitic and Saprophytic Fungi on Lichens

Lichenicolous fungi are a highly specialized ecological group of fungi obligately to facultatively growing on lichens. Most species are restricted to a particular host lichen genus. There are about 1,300 known species in 300 genera, which outnumbers the Myxomycota—Slime Molds (ca. 700 spp.)—or Ustomycetes—Smut Fungi (ca. 1,100 spp.)—and comprises about 10 percent of the known lichen species diversity. Most lichenicolous fungi belong to the Ascomycota or anamorphic fungi, but there are also over 50 species of primitive lichenicolous basidiomycota. Most lichenicolous fungi are commensals and do not obviously damage or benefit their hosts, though some are gall-inducing, some are saprobes, and some are heavy pathogens, such as the lichenicolous basidiomycete *Athelia arachnoidea* (Berk.) Jülich, which ultimately kills its host lichens.

Lichenicolous fungi (see Fig. 16o) are found worldwide where lichens are found and inhabit all groups of lichens. One particularly rich host genus is *Peltigera*, which supports over 100 species of lichenicolous fungi, a number greater than *Peltigera* species. Tundra biomes are among the richest in lichenicolous fungi, compared to other well-documented biome types. A usual proportion of lichens to lichenicolous fungi species in the best-studied Holarctic biotas (e.g., in Scandinavia or the British Isles) is about seven to one. Some lichenicolous fungi, like the pyrenomycete *Stigmidium conspurcans* (Th. Fr.) Triebel & R. Sant. found growing on squamules of *Psora rubiformis* (Ach.) Hook., are extremely common on their hosts. But they are easily overlooked, and special training is needed to set your eye to detect them, which is why they don't have common names. In many cases, lichenicolous fungi look like what are suspected to be lichen "fruits." This has resulted in many strange descriptions. For instance, the sterile crustose lichen *Lepraria neglecta* (Nyl.) Erichsen was originally described as *Lecidea neglecta* Nyl. due to the presence of numerous apothecia that actually belonged to the lichenicolous ascomycete *Llimoniella neglecta* (Vain.) Triebel & Rambold. Another example is found in the description of "*Cetraria capitata* Lynge," a sorediate "lichen" that was really only *Cetraria nigricans* Nyl. infected by the lichenicolous hyphomycetes, *Taeniolella rolfii* Diederich & Zhurb., which gave a sorediate appearance to the host thallus.

In spite of much recent work on lichenicolous fungi, many new species remain to be discovered. In Alaska there are fewer than 50 species of lichenicolous fungi known, whereas there are suspected to be no fewer than 300! Every year within the last decade, tens of new species of lichenicolous fungi have been described worldwide, including those from the Alaska interior, where paratypes of *Odontotrema melaneliae* Diederich & Zhurb. and *Odontotrema ochrolechiae* Diederich, Holien, Zhurb. & Etayo were recently found.

FAMILY: Genus of uncertain position within the Ascomycota

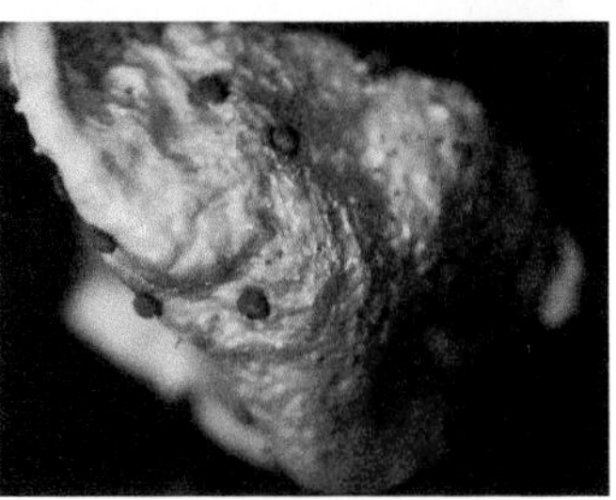

Figure 150a–b. *Abrothallus parmeliarum* (Sommerf.) Arnold (1869)

GENUS AND SPECIES: *Abrothallus parmeliarum* (Sommerf.) Nyl. (1869) (Fig. 150a–b)

STRIKING FIELD CHARACTERS: Conspicuously black (often with a greenish tinge), subglobose, constricted sessile ascomata at the base of lichen sporocarps. Often associated with another gall-inducing lichenicolous fungus, *Phacopsis oxyspora* (see below).

MACRO AND MICRO DESCRIPTIONS: Vegetative mycelium is inconspicuous. Apothecia are superficial, strongly convex, 0.1–0.4 (–0.5) mm diameter, black, with green pruina. Epihymenium is dark olive-brown. Hymenium colorless to pale green. Hypothecium is dark reddish brown. Asci are eight-spored. Ascospores are dark brown, one septate, 11.5–16 (–17.5) x 4.5–7 µm.

HABITAT AND ROLE: The fungus is not rare on thalli and apothecia of *Parmelia* s. str. (in the strict sense) species, including *P. omphalodes* (L.) Ach. It is parasymbiotic as it causes no visible damage to its host lichen.

FAMILY: Arthoniaceae

GENUS AND SPECIES: *Arthonia clemens* (Tul.) Th. Fr. (1867) (Fig. 151)

STRIKING FIELD CHARACTERS: Seen as large, conspicuously black, and slightly convexed dots (apothecia) immersed in the *Rhizoplaca* host hymenium.

Figure 151. *Arthonia clemens* (Tul.) Th. Fr. (1867)

MACRO AND MICRO DESCRIPTIONS: Vegetative mycelium is inconspicuous. Apothecia are spotlike, erumpent, flat to slightly convex, immarginate, extending up to several millimeters, and blackish. Epihymenium is brown. Hypothecium is colorless. Asci are eight-spored. Ascospores are colorless, soleiform (shaped like the sole of a shoe with one cell larger than the other), one septate, 9–15 x 3–6 µm.

HABITAT AND ROLE: This fungus grows in the hymenium of its host's apothecia. It is often found on *Rhizoplaca* species as a pathogen.

FAMILY: Arthoniaceae

GENUS AND SPECIES: *Arthonia stereocaulina* (Ohlert) R. Sant. (1993) (Fig. 152)

STRIKING FIELD CHARACTERS: Small (use hand lens!), black irregular-convexed dots (apothecia) form over the phyllocladia of the host. *Stereocaulon.*

MACRO AND MICRO DESCRIPTIONS: Vegetative mycelium is inconspicuous. Apothecia soon become superficial, subglobose, or irregular in shape, (0.05–) 0.1–0.25 (–0.5) mm diameter, black, shiny, arising singly, or aggregated and/or confluent. Epihymenium and hymenium are olivaceous brown. Hypothecium is brown. Asci are eight-spored. Ascospores are colorless, obpyriform, one-septate, (10.5–) 11.5–13–14 (–16) x 4–7.5 µm.

Figure 152. *Arthonia stereocaulina* (Ohlert) R. Sant. (1993)

HABITAT AND ROLE: This rather rare and inconspicuous parasymbiotic fungus grows on phyllocladia of *Stereocaulon alpinum* Lauer ex. Funck, as well as on many other *Stereocaulon* species.

FAMILY: A genus of uncertain position within the Dothideomycetes.

GENUS AND SPECIES: *Cercidospora stereocaulorum* (Arnold) Hafellner (1987) (Fig. 153)

Figure 153. *Cercidospora stereocaulorum* (Arnold) Hafellner (1987)

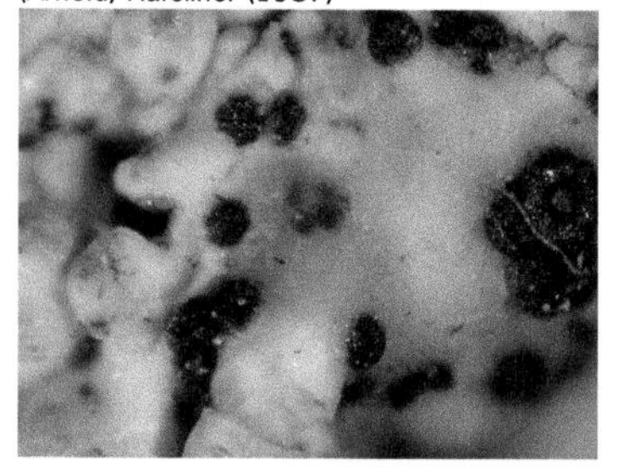

STRIKING FIELD CHARACTERS: The fungus appears as small black jugs or pitchers, (perithecia) with a wide opening (use a hand lens!) on the branch of the *Stereocaulon* host.

MACRO AND MICRO DESCRIPTIONS: Vegetative mycelium is inconspicuous. Perithecia are subglobose, often applanate from above and with a crateriform depression around the ostiole, 0.1–0.2 (–0.3) mm diameter, black, shiny, usually one-third to one-half emergent, arising singly or more often aggregated into groups. Peridium is emerald green in its upper part. Paraphyses are branched. Asci are (2–) 4 (–8)-spored. Ascospores are colorless, boat-shaped, usually with attenuated "tails," (1–) 3 (–6)-transseptate, (15–) 19–25.5 (–30) x (4–) 5.5–7 (–8) µm.

HABITAT AND ROLE: This rather common parasymbiotic fungus grows on various *Stereocaulon* species, including *S. glareosum* (Savicz) H. Magn.

FAMILY: A genus of uncertain position within the Lecanoromycetes

GENUS AND SPECIES: *Corticifraga peltigerae* (Fuckel) D. Hawksw. & R. Sant. (1990) (Fig. 154)

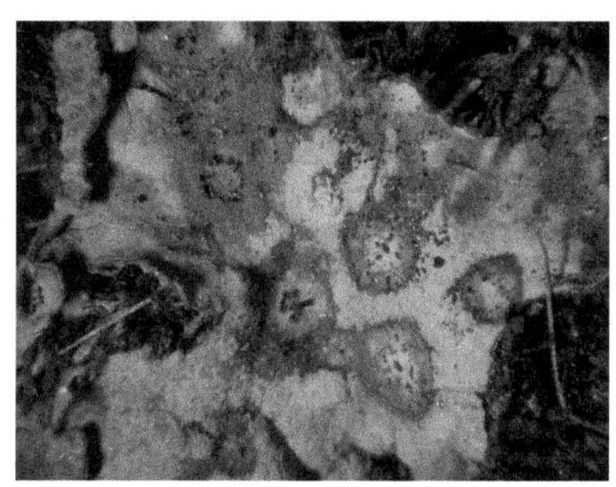

Figure 154. *Corticifraga peltigerae* (Fuckel) D. Hawksw. & R. Sant. (1990)

STRIKING FIELD CHARACTERS: This fungus is conspicuous, dark, appearing as concave dots, coins, or sconelike apothecia of angular shapes, sometimes with torn margins, associated with bleached necrotic circles on the lobes of the *Peltigera* or *Solorina* host.

MACRO AND MICRO DESCRIPTIONS: Vegetative mycelium is inconspicuous. Apothecia are erumpent through the host cortex, immersed, or only the disc exposed, concave to plane, 0.2–0.4 mm diameter, brown to black, and arising in circular groups in bleached necrotic patches. Exciple is well developed in young apothecia, becoming almost absent when mature, not or scarcely extending above the surface of the disc, dark olivaceous brown. Paraphyses are filiform with capitate, brown apices. Asci eight-spored. Ascospores colorless, ellipsoid, (1–) 3-septate, 20–23 x 5–7 µm.

HABITAT AND ROLE: This is a rather common pathogenic fungus on thalli of various *Peltigera* species, including *P. rufescens* (Weiss) Humb. (photo). It is also known from *Solorina crocea* (L.) Ach. and causes bleaching of the host thalli.

FAMILY: Odontotremataceae

GENUS AND SPECIES: *Geltingia associata* (Th. Fr.) Alstrup & D. Hawksw. (1990) (Fig. 155)

Figure 155. *Geltingia associata* (Th. Fr.) Alstrup & D. Hawksw. (1990)

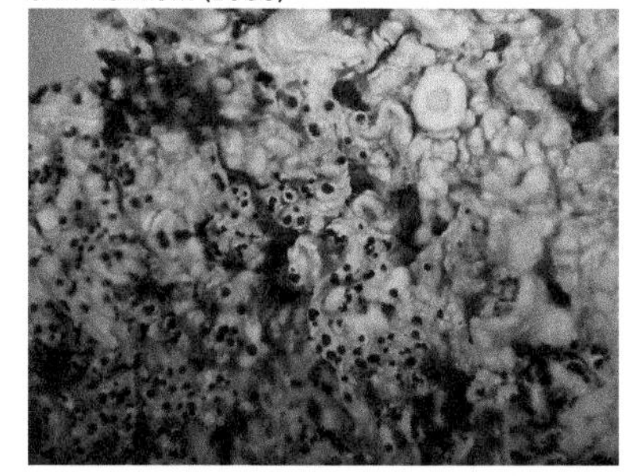

STRIKING FIELD CHARACTERS: This fungus appears as conspicuous black dots, coins, or sconelike apothecia, which are craterlike at first, then become flat and protrude. Infrequently numerous.

MACRO AND MICRO DESCRIPTIONS: Vegetative mycelium is inconspicuous. Apothecia erumpent through the *Ochrolechia* host cortex, orbicular, usually with plane disk and prominent margin, black, 0.1–0.4 mm diameter. Exciple and hypothecium brown. Hymenium colorless. Paraphyses filamentous, simple, not swollen apically. Asci eight-spored. Ascospores are colorless, subglobose, nonseptate, 5–9 μm diameter, uniseriate in an ascus.

HABITAT AND ROLE: This parasymbiotic fungus grows on thalli of *Ochrolechia* species, including *O. frigida* (Sw.) Lynge.

FAMILY: Uncertain, belongs within the Mitosporic fungi (Deuteromycetes)

GENUS AND SPECIES: *Illosporium carneum* Fr. (1829) (Fig. 156)

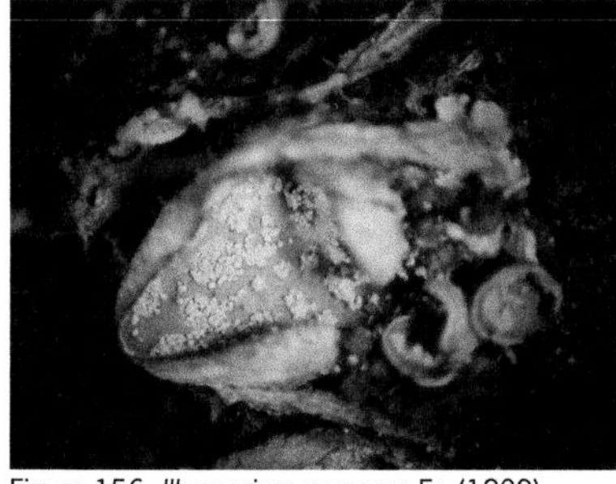

Figure 156. *Illosporium carneum* Fr. (1829)

STRIKING FIELD CHARACTERS: Occurs as conspicuous bright pink globose heaps, composed of discrete large grains (use a hand lens!) on the *Peltigera* host lobes.

MACRO AND MICRO DESCRIPTIONS: Sporodochia (small, cushion-shaped stroma bearing short conidiophores that produce the spores, or cinidia) are delimited, densely compacted, irregularly applanate (flattened), 0.2–0.3 mm diameter, and pale pink. Conidia are colorless singly, but pale pink in mass, subglobose and nonseptate, 6–7 μm diameter, often adhering in compact masses ("grains") up to 0.1 mm diameter.

HABITAT AND ROLE: This remarkable parasymbiotic hyphomycete is not rare on thalli of *Peltigera didactyla* (With.) J. R. Laundon and some other *Peltigera* species.

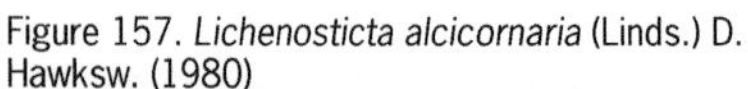

Figure 157. *Lichenosticta alcicornaria* (Linds.) D. Hawksw. (1980)

FAMILY: Uncertain, belongs within the Mitosporic fungi (Deuteromycetes)

GENUS AND SPECIES: *Lichenosticta alcicornaria* (Linds.) D. Hawksw. (1980) (Fig. 157)

STRIKING FIELD CHARACTERS: Rather conspicuous; appear as numerous dark dots, coins, or scone-like pycnidia, which are typically most abundant on the underside of the *Cladonia* host squamules.

MACRO AND MICRO DESCRIPTIONS: Vegetative mycelium is inconspicuous. Pycnidia semi-immersed, subglobose to broadly pyriform, about 0.1 mm diameter, brown to black, scattered and often numerous. Conidia colorless, lacriform, with rounded apex and attenuated base, noonseptate, 6–10 x 3–5 µm.

HABITAT AND ROLE: This parasymbiotic fungus grows on many *Cladonia* species, mainly on the underside of their squamules.

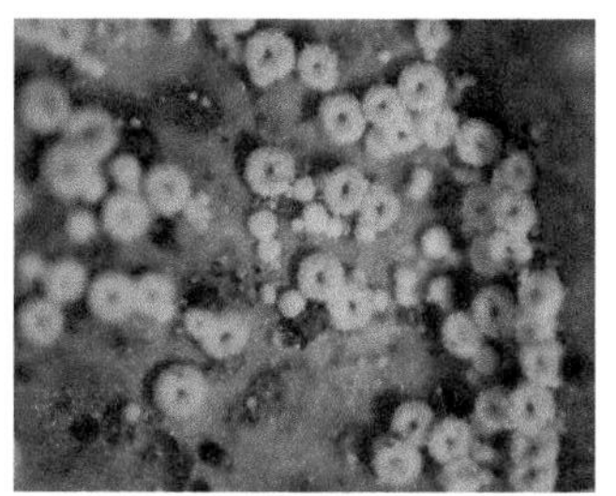

Figure 158. *Nectriopsis lecanodes* (Ces.) Diederich & Schroers (1999)

FAMILY: Bionectriaceae

GENUS AND SPECIES: *Nectriopsis lecanodes* (Ces.) Diederich & Schroers (1999) (Fig. 158)

STRIKING FIELD CHARACTERS: Occurs as conspicuously aggregated cream-colored, hairy (use a hand lens!) cups (perithecia) with inrolled margins and wide openings, sitting on the lichen host lobes. Perithecia often numerous.

MACRO AND MICRO DESCRIPTIONS: Vegetative mycelium inconspicuous. Perithecia superficial, pale pink or orange, 0.15–0.3 mm diameter, and usually scattered. Peridium hyaline to pale brown. Asci eight-spored. Ascospores colorless, ellipsoid, one–septate, 8–12 x 3–4.5 µm.

HABITAT AND ROLE: This parasymbiotic fungus is not rare on thalli of various *Peltigera* species, as well as on *Nephroma, Lobaria,* and some other lichen genera.

Figure 159. *Nesolechia cetrariicola* (Linds.) Arnold (1874)

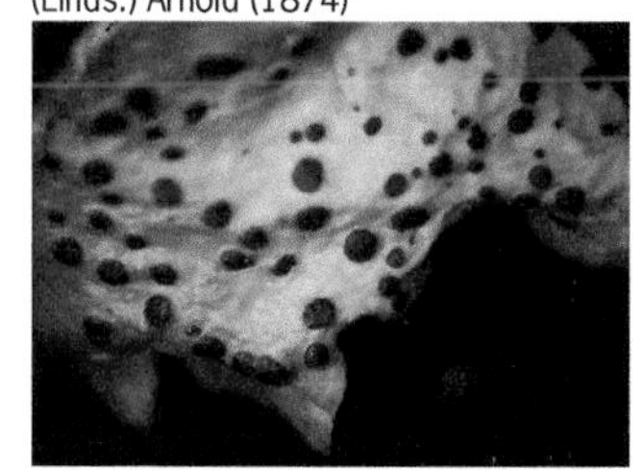

FAMILY: Parmeliaceae

GENUS AND SPECIES: *Nesolechia cetrariicola* (Linds.) Arnold (1874) (Fig. 159)

STRIKING FIELD CHARACTERS: Occurs as conspicuously aggregated dark brown, rounded, flat to slightly convex dots, coins, or scones (apothecia) on the *Cetraria* host lobes.

MACRO AND MICRO DESCRIPTIONS: Vegetative mycelium inconspicuous. Apothecia superficial, convex to subglobose, 0.2–0.5 mm diameter, dark brown to almost black, and glossy. Epihymenium and hypothecium dark brown. Hymenium brown. Paraphyses are with swollen, dark brown apices. Asci eight-spored. Ascospores fusiform, colorless, 0(–1)-septate, 9–15 x 3–4 µm.

HABITAT AND ROLE: This fungus grows on thalli of *Cetraria islandica* (L.) Ach. and some other *Cetraria* species. It can induce gall formation, but otherwise causes no visible damage to the host lichen.

Figure 160. *Phacopsis oxyspora* (Tul.) Triebel & Rambold (1988)

FAMILY: Parmeliaceae

GENUS AND SPECIES: *Phacopsis oxyspora* (Tul.) Triebel & Rambold (1988) (Fig. 160)

STRIKING FIELD CHARACTERS: Apothecia are conspicuous aggregated, dark brown, rounded, flat to slightly convex clots, coins, or scones on the lobes of its *Parmelia* host, infrequently associated with swellings on the host. It is often associated with another lichenicolous fungus, *Abrothallus parmeliarum* (see above).

MACRO AND MICRO DESCRIPTIONS: Vegetative mycelium inconspicuous. Apothecia immersed to sessile, circular, immarginate (with no well-defined edge), flat to convex, dark brown or black. Epihymenium and hymenium colorless to brown. Paraphyses branched, with thickened brown apices. Hypothecium colorless to dark brown. Asci eight-spored. Ascospores colorless, nonseptate, ellipsoid, and usually with attenuated ends, (11–) 14–18.5 (–24) x (5–) 5.5–6.5 (–7) µm.

HABITAT AND ROLE: This is a common fungus growing on thalli of *Parmelia omphalodes* (L.) Ach. and many other parmelioid lichens. Often it induces gall-like swellings of the host thalli, but otherwise causes no visible damage to it.

Figure 161. *Polycoccum bryonthae* (Arnold) Vězda (1969)

FAMILY: Dacampiaceae

GENUS AND SPECIES: *Polycoccum bryonthae* (Arnold) Vězda (1969) (Fig. 161)

STRIKING FIELD CHARACTERS: Infected apothecia of the lichen host become noticeably blackish and are filled with black jugs or pitcherlike fruitbodies (perithecia), and often crowded together.

MACRO AND MICRO DESCRIPTIONS: Vegetative mycelium inconspicuous. Perithecia semi-immersed, subglobose, about 0.1 mm diameter, blackish. Asci eight-spored. Ascospores olive brown, soleiform, 1 (–2)-septate, 11–13 x 4–6 µm, and uniseriate in an ascus.

HABITAT AND ROLE: This strongly pathogenic fungus grows in apothecia of terricolous *Caloplaca*, *Lecanora*, and *Pertusaria* species, where it kills and destroys the host's hymenium. The illustration (Fig. 163) is on *Lecanora epibryon* (Ach.) Ach.

Figure 162. *Polycoccum trypethelioides* (Th. Fr.) R. Sant. (1960)

FAMILY: Dacampiaceae

GENUS AND SPECIES: *Polycoccum trypethelioides* (Th. Fr.) R. Sant. (1960) (Fig. 162)

STRIKING FIELD CHARACTERS: Occurs as conspicuously large swellings or bulbs on the Stereocaulon host branches with numerous black "eyes" or speckles which are protruding upper portions of the fungus sporocarps (perithecia)

MACRO AND MICRO DESCRIPTIONS: Vegetative mycelium inconspicuous. Perithecia subglobose to broadly obpyriform, ostiolate, (0.1) 0.5–0.3 (–0.6) mm diameter, black, glossy, almost completely immersed to occasionally up to one-half emergent when old. Tightly aggregated on bullate galls on the stems of the host thallus (young galls concolorous with host thallus, later often light brown, and occasionally fawn brown or cinnamon brown). Up to 200 perithecia per gall, adjacent to confluent (running into one another). Peridium brown. Interascal filaments branched, anastomosed. Asci (4–) eight-spored. Ascospores obpyriform to obovate at first hyaline then olive to brown, (0–) one–septate, (12–) 15–19 (–22.5) x (7–) 8.5–10.5 (–12.5) µm.

HABITAT AND ROLE: This fungus grows on branches of *Stereocaulon alpinum* and many other *Stereocaulon* species, inducing gall-like swellings, but otherwise causes no damage to the host. It is infrequent, but often abundant.

Figure 163. *Rhagadostoma lichenicola* (De Not.) Keissl. (1930)

FAMILY: Nitschkiaceae

GENUS AND SPECIES: *Rhagadostoma lichenicola* (De Not.) Keissl. (1930) (Fig. 163)

STRIKING FIELD CHARACTERS: Perithecia are noticeably robust, black, crowded to congested, superficial jugs, or pitcherlike fruitbodies with a rugose (wrinkled) (use a hand lens!) surface.

MACRO AND MICRO DESCRIPTIONS: Vegetative mycelium inconspicuous. Perithecia superficial, ovoid, with irregular opening, 0.2–0.5 mm diameter, black, with coarsely verrucose wall, and often densely crowded. Exciple black and thick. Asci (2–) four-spored. Ascospores colorless, fusiform, (0–) 1 (–3)-septate, 30–50 x 7–10 µm.

HABITAT AND ROLE: This common and conspicuous fungus grows on *Solorina* and *Peltigera* species, mostly on *S. crocea* thalli. It is often abundant and pathogenic in heavy infections.

FAMILY: Pilocarpaceae

GENUS AND SPECIES: *Scutula stereocaulorum* (Anzi) Körb. (1865) (Fig. 164)

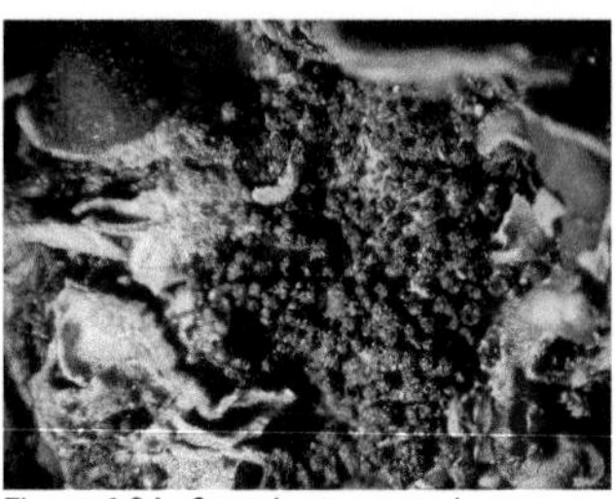

Figure 164. *Scutula stereocaulorum* (Anzi) Körb. (1865)

STRIKING FIELD CHARACTERS: Apothecia are conspicuously numerous, and are aggregated to crowded, as large brown, rounded convex dots, coins, or scones on the *Stereocaulon* host phyllocladia or branches.

MACRO AND MICRO DESCRIPTIONS: Vegetative mycelium inconspicuous. Apothecia 0.1–0.8 mm diameter, blackish brown, glossy, sessile, flat to slightly convex, and constricted at the base. Epihymenium dark reddish brown. Paraphyses capitate, with brown caps. Asci eight-spored. Ascospores colorless, fusiform, (0–)one–septate, (10.5–) 13.5–17.5 (–19) x (3.5–) 4–4.5–5.5 (–6.5) µm.

HABITAT AND ROLE: This is one of the most common and abundant lichenicolous fungi, growing parasymbiotically on phyllocladia of various *Stereocaulon* species, including *S. alpinum* (Fig. 148).

FAMILY: Mycosphaerellaceae

GENUS AND SPECIES: *Sphaerellothecium araneosum* (Arnold) Zopf n.d. (Fig. 165)

Figure 165. *Sphaerellothecium araneosum* (Arnold) Zopf (RSD)

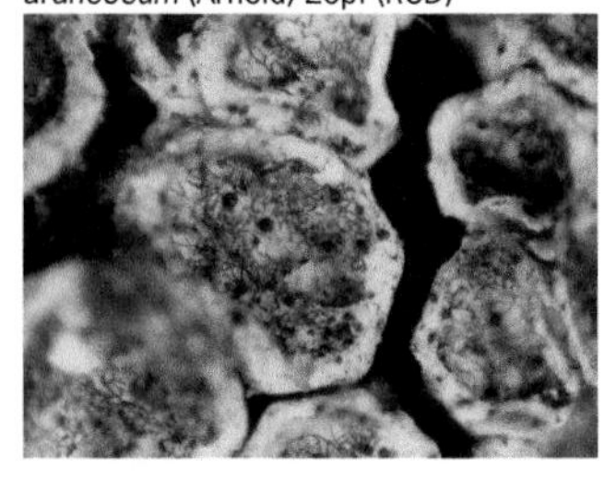

STRIKING FIELD CHARACTERS: Produces a characteristic superficial net of black reticulate hyphae with incorporated small blackish brown dots (perithecia) on the *Ochrolechia* host's surface.

MACRO AND MICRO DESCRIPTIONS: Vegetative hyphae conspicuous, superficial, forming a sparsely branched, dark brown reticulum. Perithecia sessile, blackish brown, subglobose, 30–50 µm diameter. Asci eight-spored, a few per ascomata. Ascospores colorless, lageniform or lecythiform (skittle-shaped), one–septate, 9–12 x 4–5 µm.

HABITAT AND ROLE: The fungus is not rare on apothecia and thalli of *Ochrolechia* species, including *O. upsaliensis* (L.) A. Massal., as well as on some other lichen genera. It can be pathogenic in heavy infections.

FAMILY: Mycosphaerellaceae

GENUS AND SPECIES: *Sphaerellothecium minutum* Hafellner (1993) (Fig. 166)

Figure 166. *Sphaerellothecium minutum* Hafellner (1933)

STRIKING FIELD CHARACTERS: No other known fungus produces numerous dark, small (use a hand lens!) dots or speckled perithecia on thalli of the host genus *Sphaerophorus*.

MACRO AND MICRO DESCRIPTIONS: Vegetative hyphae discernible on the host's branches, forming a dark brown reticulum. Perithecia semi-immersed, subglobose, dark brown, and 60–80 µm diameter. Peridium dark brown. Asci eight-spored, a few per ascomata. Ascospores colorless, soleiform, and one–septate, 9–13 x 3–5 µm.

HABITAT AND ROLE: The fungus is very common on thalli of *Sphaerophorus globosus* (Huds.) Vain. and *S. fragilis* (L.) Pers. It is usually abundant, but causes no visible damage to the hosts.

Figure 167. *Sphinctrina turbinata* (Pers.) De Not. (1846)

FAMILY: Sphinctrinaceae

GENUS AND SPECIES: *Sphinctrina turbinata* (Pers.) De Not. (1846) (Fig. 167)

STRIKING FIELD CHARACTERS: No other known fungi on the host *Pertusaria* species have these characteristic stipitate club-shaped to clavate (use a hand lens!) and rather conspicuous black sporocarps (apothecia).

MACRO AND MICRO DESCRIPTIONS: Vegetative mycelia inconspicuous. Apothecia clavate, composed of a short stalk and capitulum, 0.2–0.7 mm high, 0.2–0.5 mm diameter, dark brown to black with black extruding mazaedium (mature spores become free from the asci as a dry powdery mass) above and usually single. Exciple dark red-brown. Hypothecium colorless. Asci eight-spored. Ascospores dark brown, subglobose to cuboid, nonseptate, 3.5–7 x 4.5–7 µm, and verrucose.

HABITAT AND ROLE: This parasymbiotic fungus mostly colonizes thalli of *Pertusaria* species. It is rare and never abundant.

Figure 168. *Stigmidium conspurcans* (Th. Fr.) Triebel & R. Sant. (1989)

FAMILY: Mycosphaerellaceae

GENUS AND SPECIES: *Stigmidium conspurcans* (Th. Fr.) Triebel & R. Sant. (1989) (Fig. 168)

STRIKING FIELD CHARACTERS: No other fungi produce these numerous conspicuous black dots or speckles (perithecia) on the squamules of its host, *Psora rubiformis*, which cause no discoloration or other visible damage.

MACRO AND MICRO DESCRIPTIONS: Vegetative mycelium inconspicuous. Perithecia semi-immersed, subglobose, about 0.1 mm diameter, and dark brown. Peridium dark brown above, pale brown at base. Asci eight-spored. Ascospores colorless, soleiform, one–septate, and 12–15 x 4–6 µm.

HABITAT AND ROLE: The fungus is a faithful companion of *Psora rubiformis* (Ach.) Hook., where it is very common and usually abundant. It causes no visible damage to the host, although it can disrupt its squamules with heavy infections.

FAMILY: Uncertain, belongs within the Mitosporic fungi (Deuteromycetes)

GENUS AND SPECIES: *Taeniolella beschiana* Diederich (1992) (Fig. 169)

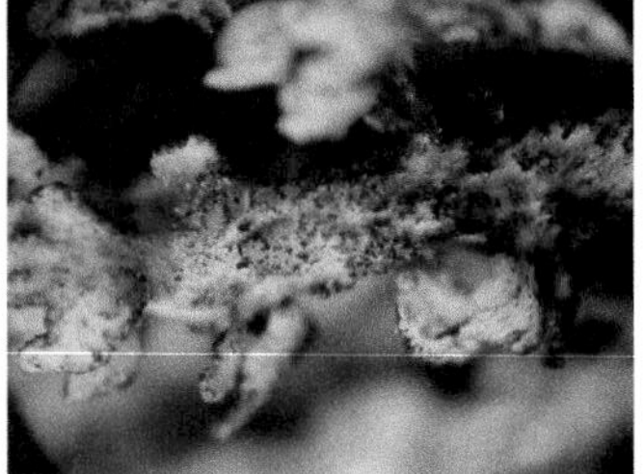

Figure 169. *Taeniolella beschiana* Diederich (1992)

STRIKING FIELD CHARACTERS: These fungi exist as tiny (use a hand lens!), dense, dark bristles spread over the *Cladonia* host thalli.

MACRO AND MICRO DESCRIPTIONS: Colonies effused, dark brown to black, and dispersed. Conidiophores erect, branched below, dark brown, and 40–50 x 3–6 µm. Conidia brown, single or catenate (chainlike) by 2, ellipsoid, truncated at both ends and zero to one–septate, 6–10 x 3–5 µm.

HABITAT AND ROLE: This parasymbiotic fungus grows on thalli of *Cladonia stricta* (Nyl.) Nyl. and on some other *Cladonia* species, where it is most abundant on their squamules.

FAMILY: Uncertain, belongs within the Mitosporic fungi (Deuteromycetes)

GENUS AND SPECIES: *Taeniolella pertusariicola* D. Hawksw. & H. Mayrhofer (1990) (Fig. 170)

Figure 170. *Taeniolella pertusariicola* D. Hawksw. & H. Mayrhofer (1990)

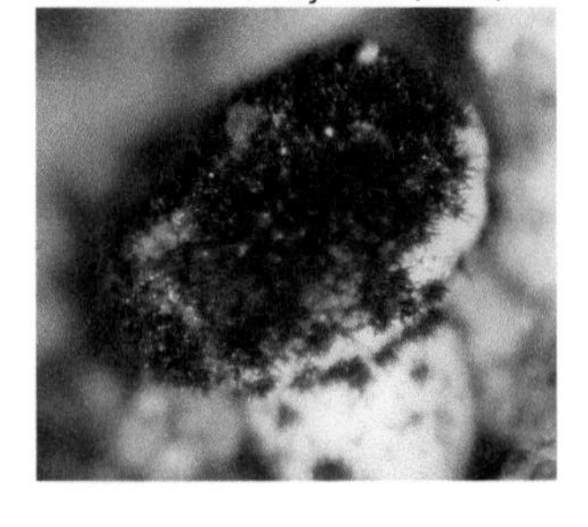

STRIKING FIELD CHARACTERS: Exists as tiny (use a hand lens!), dense, dark bristles on the host apothecia. Infected apothecia are conspicuous, becoming blackish when compared with noninfected hosts.

MACRO AND MICRO DESCRIPTIONS: Colonies superficial, blackish, scattered or loosely aggregated. Conidiophores forming caespitose tufts, erect, mostly unbranched, dark brown, verrucose, and 10–30 x 6–8 μm. Conidia dark brown, broadly ellipsoid to dolioform, 1–3-septate, verrucose, and 11–16 x 6–8 μm.

HABITAT AND ROLE: This fungus typically grows on the hymenium of apothecia on *Pertusaria*, including *P. bryontha* (Ach.) Nyl., and *Varicellaria* species. It is somewhat pathogenic in dense infections.

FAMILY: Thelocarpaceae

GENUS AND SPECIES: *Thelocarpon epibolum* Nyl. (1866) (Fig. 171)

STRIKING FIELD CHARACTERS: No other known fungi produce these tiny (use a hand lens!), pale green, semi-immersed juglike or pitcherlike perithecia on the upper lobe surfaces of its host, *Peltigera aphthosa*.

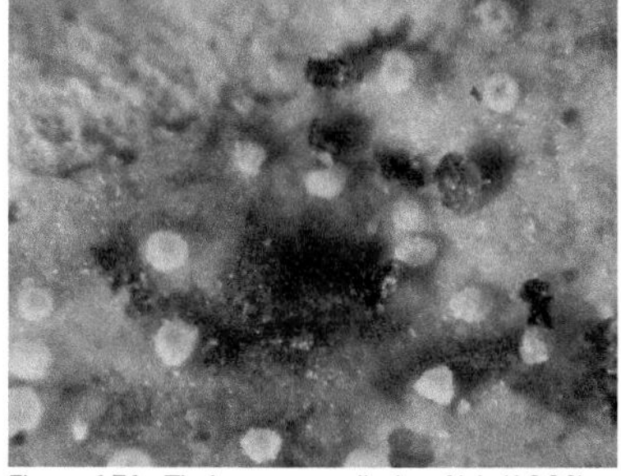
Figure 171. *Thelocarpon epibolum* Nyl. (1866)

MACRO AND MICRO DESCRIPTIONS: Thallus of inconspicuous, sometimes lichenized warts. Apothecia superficial, perithecium-like, ovoid, 0.2 mm high, 0.1 mm diameter, pale yellow and pruinose. Peridium colorless. Paraphyses thin, simple. Asci flask-shaped, 50–300-spored. Ascospores colorless, oblong, nonseptate, and 4–7 x 2–3 μm.

HABITAT AND ROLE: This tiny lichenicolous lichen grows on thalli of various lichen genera, and preferably on *Peltigera*. It is parasymbiotic or saprophytic.

Lichenized Fungi: The Lichens

One might wonder why lichens have been included in a mushroom field guide. Lichens, which are very common throughout interior Alaska, are partly fungi (see Fig. 16p). In fact, the lichen body, the **thallus**, is composed mostly of fungal hyphae interspersed with photosynthetic green algae or cyanobacteria or both. Thus, the lichen is what might be termed a "composite" organism, since it consists of two (or even three) entirely different organisms in a close symbiotic or mutually beneficial relationship. The photosynthetic algal or cyanobacterium partner in this association is the **photobiont**, while the fungal partner is referred to as the **mycobiont**. The photobionts are members of two entirely different kingdoms. The green algae belong to the phototrophic eukaryotic Protista (Protoctista) and the cyanobacteria to the phototrophic prokaryotic Monera. As noted in the introduction, fungi belong to the Kingdom Fungi (or Myceteae), as do the lichens.

The association between the photobiont and mycobiont is beneficial to both; the fungus captures water, essential minerals, and some vitamins for the photobiont, while the photobiont provides photosynthetic products to the fungus. The fungus provides structural support for the photobiont and a favorable microclimatic and chemical environment in which the photobiont can grow and flourish, and also serves as an effective filter of excess photosynthetically active and ultraviolet radiation.

The vast majority of lichenized fungi produce ascospores; hence, they are **ascolichens** (approximately 99 percent of all lichens) and typically have disc-shaped spore-producing surfaces. The group includes crustose, foliose, fruticose, and squamulose forms. The remaining examples of lichenized fungi produce basidiospores and thus are **basidiolichens** (less than 1 percent of all lichens). The basidiospores emanate from upright and clubbed coral fungi or gilled agaric fruitbodies. In each instance, the fungus partner determines the shape of the lichen thallus. However, the free-living fungus does not assume the morphological form expressed in the lichen unless combined with its photosynthetic partner.

Lichens generally fall into three categories: crustose, foliose, and fruticose. **Crustose lichens**, as their name implies, grow very closely attached to their substrate. Structurally, they consist of an upper differentiated cortex layer that is usually pigmented; an algal layer seen as a green or bluish green layer just below the cortex; and a medulla, composed of loosely arranged fungal hyphae that also serve to attach the lichen to the substrate. **Foliose lichens**, as their name suggests, are flattened with often leaflike lobes. Structurally, they have an upper cortex, an algal layer immediately below, a medulla of loosely arranged fungal hyphae, and a differentiated lower cortex. Foliose lichens may be closely attached to the substrate by wrinkles or folds in the lower cortex, by special hair- or fingerlike projections or rhizines on the lower surface, or, as found in *Umbilicaria* species, by a central umbilicus. Some lichens, such as the reindeer antler lichen *Masonhalea*, are not attached to the substrate and can be blown around freely by wind. **Fruticose lichens** often grow erect on their substrate or, if growing on twigs and branches, can form pendent masses. Structurally, they have an outer cortical layer, an algal layer below this, and an inner medullary layer. Some fruticose lichens, such as *Usnea*, have a differentiated stiff central strand, which gives the lichen considerable strength and flexibility.

The number of species of lichens found worldwide is estimated to be between 13,500 and 17,000. Although rich with assorted lichens, the Alaska interior has far fewer species than this. The examples included here are among the more common lichens one is likely to encounter on a casual walk through forest or tundra landscapes. Cold-tolerant lichens often dominate higher subarctic and alpine environments. Competition from higher plants is significantly reduced in these cold-stressed regions and lichens are able to spread luxuriously over organic and mineral soils—found most often among mosses; on rocks, bark, and wood; on old antlers and bones; and under intermittent snowbanks and sometimes in quite wet environments, although lichens generally do not care for extreme wet habitats.

Color reactions with various added chemical solutions are important as an aid to the identification of many lichen species. Commonly applied to parts of a lichen thallus, particularly the medulla layer, are solutions of potassium hydroxide (abbreviated as K in the following section). A color response is designated as K+. If there is no color reaction, the response is given as K-. Also applied to the medulla is a solution of calcium hypochlorite (abbreviated as C in the following section). The hypochlorite solution may also be applied immediately after application of the potassium hydroxide solution (abbreviated as KC). A solution of a hazardous organic chemical, paraphenylene diamine, is also sometimes used (abbreviated as P). A solution of iodine with potassium iodide is also used to examine the fine structural differences of the fruiting structure (apothecium) and also the structural detail of the tip of the spore-bearing ascus. Some species of lichens may also fluoresce under short-wave ultraviolet radiation (abbreviated as UV) at 254 nanometers wavelength, and this may also be used as an aid in species identification.

Ascolichens: Crustose

FAMILY: Icmadophilaceae

GENUS AND SPECIES: *Icmadophila ericetorum* (L.) Zahlbr. (1895) (Fig. 172a–d)

COMMON NAME: Candy, spraypaint, fairy, or peppermint drop lichen

STRIKING FIELD CHARACTERS: **Thallus** consisting of rounded pink to grayish pink discs (apothecia) on a smooth to granular crustose bed that is off-white to pale

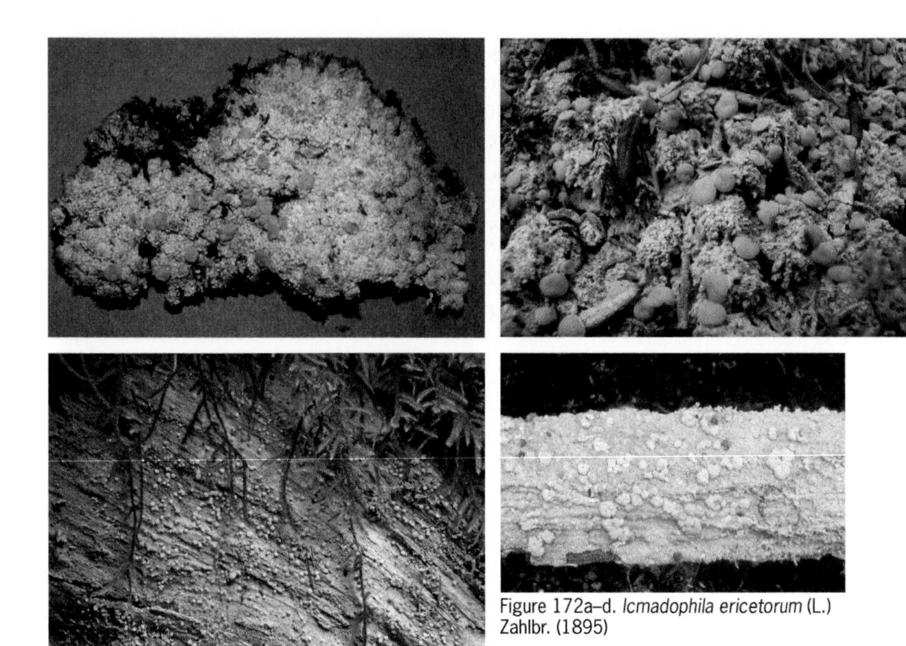

Figure 172a–d. *Icmadophila ericetorum* (L.) Zahlbr. (1895)

green and on old conifer logs, stumps, or barren soil. Sometimes also found on old compact *Sphagnum* hummocks or cushion-forming plants.

MACRO AND MICRO DESCRIPTION: Thallus pink to grayish pink, granular; apothecia pale red, margin thick but becoming excluded when older. Thallus K+ yellow, apothecia K+ orange, C+ orange. Spores fusiform, two to four celled, 15–27 x 4–6 µm.

HABITAT AND ROLE: Common and widespread, occurring on shaded rotting spruce logs, soil, humus, and mosses, especially some species of *Sphagnum*. In southern parts of Alaska, *Icmadophila ericetorum* is also found on tree trunks over bark or wood.

Ascolichens: Foliose

FAMILY: Nephromataceae

GENUS AND SPECIES: *Nephroma expallidum* (Nyl.) Nyl. (1865) (Fig. 173a–b)

COMMON NAME: Alpine paw or paw lichen

STRIKING FIELD CHARACTERS: **Thallus** greenish brown to brown, greener when wet, with lobes up to 1.8 cm wide. Underside light brown at margins, darker towards the center, covered with short, dense hairs and thus feltlike. Discs light to dark brown,

Figure 173a-b. *Nephroma expallidum* (Nyl.) Nyl. (1865)

up to 10 mm wide, occurring on the lower surface of lobe tips. Thallus with internal pockets (cephalodia) containing nitrogen-fixing cyanobacteria; these visible only on the lower surface.

MACRO AND MICRO DESCRIPTION: Thallus K-, apothecia K-, C-. **Spores** light brown, four-celled, 17–21 x 5–6 µm.

HABITAT AND ROLE: This lichen grows among mosses in moist subalpine spruce forests and in moist moss/heath communities in alpine areas at higher elevations throughout the Alaska interior. Another ground-dwelling paw lichen is *Nephroma arcticum* (L.) Torss. The thallus of *N. arcticum* is yellow-green and has cephalodia that are visible on the upper surface. As viewed from above, the cephalodia appear as small bluish green warts.

FAMILY: Peltigeraceae

GENUS AND SPECIES: *Peltigera aphthosa* (L.) Willd. (1787) (Fig. 174a–c)

COMMON NAME: Studded vein lichen

STRIKING FIELD CHARACTERS: **Thallus** large, leafy green with thin black, irregularly shaped warts (cephalodia) containing cyanobacteria on the upper surface.

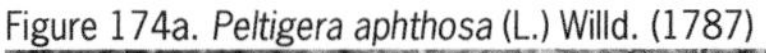

Figure 174a. *Peltigera aphthosa* (L.) Willd. (1787)

MACRO AND MICRO DESCRIPTION: K-, C-, KC-, P-. **Spores** acicular, 4–10-celled, 48–70 x 4–7 µm.

HABITAT AND ROLE: *Peltigera aphthosa* is ubiquitous throughout the Alaska interior. It is found in willow (*Salix alaxensis*) thickets, near moss-covered stream margins, in spruce forests, and even in dryer, colder heath thickets at or above treeline. *P. aphthosa* is usually found on moss-covered humic soil or on decaying wood. *P. leucophlebia* (Nul.) Gyeln. is a similar species

Figure 174b–c. *Peltigera aphthosa* (L.) Willd. (1787)

found in habitats that overlap with those occupied by *P. aphthosa*. *P. leucophlebia* is clearly veined on the lower surface and the undersides of the fertile lobe tips have a patchy green cortex. In contrast, typical specimens of *P. aphthosa* lack veins and have a continuous green cortex on the underside of each fertile lobe tip. However, the two species do intergrade in appearance. This is especially true for specimens of *P. aphthosa* that occur in habitats with a heavy snow cover. Such specimens are often difficult to distinguish from *P. leucophlebia*.

FAMILY: Peltigeraceae

GENUS AND SPECIES: *Peltigera neopolydactyla* (Gyeln.) Gyeln. (1932) (Fig. 175a–c)

COMMON NAME: Many-fingered vein lichen

STRIKING FIELD CHARACTERS: **Thallus** small to large, dark grayish-green when wet, becoming pale gray upon drying and with "crimped" edges. The chestnut-brown fruiting bodies—which occur on the tips of erect narrow, folded lobes (the "fingers")—look like fingernails.

Figure 175a–c. *Peltigera neopolydactyla* (Gyeln.) Gyeln. (1932)

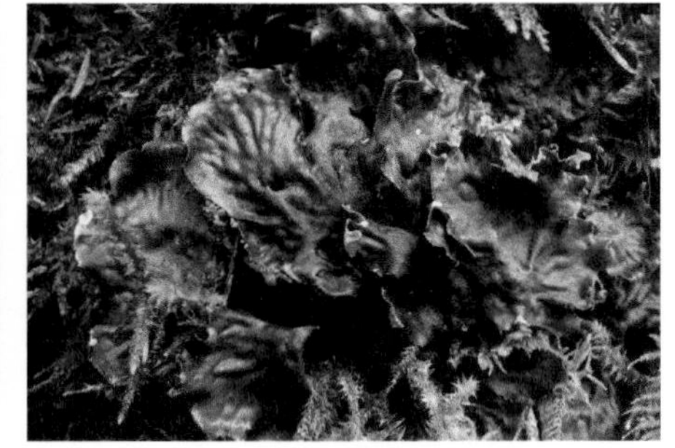

MACRO AND MICRO DESCRIPTION: K-, C-, KC-, P-. Asci amyloid. **Spores** acicular, slightly curved, 4–10-celled, 48–105 x 3–4 µm.

HABITAT AND ROLE: *Peltigera neopolydactyla* grows on soil, humus, or among mosses in moist places. the Alaska interior has a number of different species of vein lichens that are distinguished by the presence of tomentum (feltlike covering), scaly patches, or other ornamentation on the upper surface; evenness of the margins; height and width of the veins on the lower surface; and the type of rootlike attachment hairs.

FAMILY: Umbilicariaceae

GENUS AND SPECIES: *Umbilicaria hyperborea* (Ach.) Hoffm. (1796) (Fig. 176)

COMMON NAME: Northern rock tripe

STRIKING FIELD CHARACTERS: **Thallus** more or less round to somewhat lobed, attached to rocky substrates at a single point (the umbilicus). Upper surface of the thallus brown, rough, with contorted worm-shaped ridges; lower surface smooth, dark gray to brown or sometimes almost black. The concentrically fissured black circular discs on the upper surface are the fruiting bodies of the fungus.

Figure 176. *Umbilicaria hyperborea* (Ach.) Hoffm. (1796)

MACRO AND MICRO DESCRIPTION: K-, C+ red, KC+ red, P-. **Spores** simple, hyaline, ovoid, 6.6–20 x 3.3–8 µm.

HABITAT AND ROLE: This lichen grows on acid rocks and is quite common in fellfield (= feldmark) at higher elevations or on rock outcrops at lower elevations. There are at least 16 species of *Umbilicaria* in Alaska. *Umbilicaria arctica* (Ach.) Nyl. most closely resembles *U. hyperborea,* but is dove gray to tan in part on the lower surface. The ridges on the upper surface of *U. proboscidea* (L.) Schrader, another similar species, join to form a network, and there also is a raised "navel" at the attachment point.

Ascolichens: Fruticose

FAMILY: Parmeliaceae

GENUS AND SPECIES: *Flavocetraria nivalis* (L.) Kärnefelt & Thell. (1994) (Fig. 177a–b)

= *Allocetraria nivalis* (L.) Randl. & Saag

= *Cetraria nivalis* (L.) Ach.

Figure 177a–b. *Flavocetraria nivalis* (L.) Kärnefelt & A. Thell (1994)

COMMON NAME: Snow lichen

STRIKING FIELD CHARACTERS: **Thallus** surface yellow but white internally (this can be determined by breaking off a lobe and examining with a hand lens), with erect, often spreading, wrinkled, flattened, and single to clumped lobes, these reticulated (patterned/pitted). Brown discs occur on the top and sides of lobes, but are produced infrequently.

MACRO AND MICRO DESCRIPTION: K-, C-, KC+ yellow, P-. **Spores** simple, hyaline, ellipsoid, 5–8 x 3–5 µm.

HABITAT AND ROLE: This very common lichen occurs abundantly in seasonally snow-covered soil (terricolous) habitats or between clumps of mosses. Other similar yellow, ground-dwelling lichens are *Flavocetraria cucullata* (Bellardi) Kärnefelt & Thell, with inrolled margins, and *Vulpicida tilesii* (Ach.) J.E. Mattsson & M.J. Lai and *V. pinastri* (Scop.) J.E. Mattsson & M.J. Lai, which are yellow in cross section. *Vulpicida pinastri* has powdery margins, while the other species do not.

FAMILY: Parmeliaceae

GENUS AND SPECIES: *Masonhalea richardsonii* (Hook.) Kärnefelt (1977) (Fig. 178a–b)
= *Cetraria richardsonii* Hook.

COMMON NAME: Arctic tumbleweed

STRIKING FIELD CHARACTERS: **Thallus** with "silver streaks" on a smooth undersurface and a rolled-up look when dry. However, the thallus is dichotomously to irregularly divided or branched and has antlerlike, dark chestnut brown, sharp-pointed lobes with a horny texture.

Figure 178a–b. *Masonhalea richardsonii* (Hook.) Kärnefelt (1977)

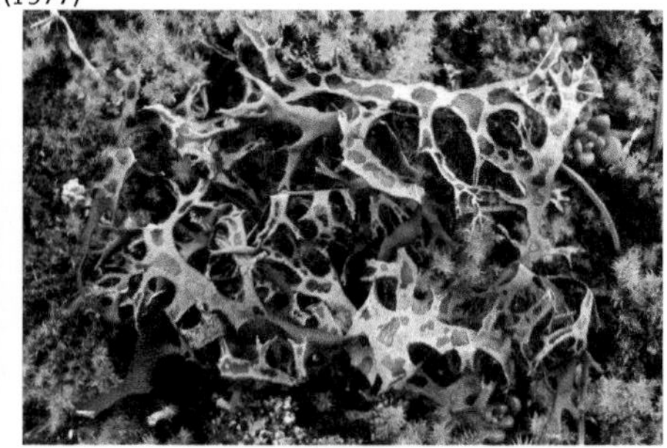

MACRO AND MICRO DESCRIPTION: K-, C+, KC+ orange-red to red, P-, UV+. **Spores** simple, hyaline, ovoid, 7.5–12.5 x 5–6 μm.

HABITAT AND ROLE: This lichen grows "unattached" and blows around on moss surfaces, often filling surface depressions.

FAMILY: Cladoniaceae

GENUS AND SPECIES: *Cladonia rangiferina* (L.) F.H. Wigg. (Fig. 179a–b)

COMMON NAME: Gray or true caribou and/or reindeer lichen

STRIKING FIELD CHARACTERS: **Thallus** gray-green, ashy gray, pale green to almost white (but never yellow-green), forming ground cover cushions or tufts, and sometimes in extensive mats. The thallus is highly branched with terminal branches in whorls of two to three around a hollow main axis, usually curved in one direction, which gives the lichen a "windswept" appearance. This species is found throughout the boreal forest.

Figure 179a–b. *Cladonia rangiferina* (L.) Nyl. (1866)

MACRO AND MICRO DESCRIPTION: K+ yellow, KC-, P+ red, UV-.

HABITAT AND ROLE: *Cladonia rangiferina* occurs on humic soils and is a source of food for the woodland caribou. It is fast growing and can form large, dense, and deep mats. *Cladina mitis* (Sandst.) Hustich and *C. arbuscula* (Warrr.) Hale & Culb. are similar in appearance, but differ in having a yellow-green color and not appearing windswept.

FAMILY: Cladoniaceae

GENUS AND SPECIES: *Cladonia stellaris* (Opiz) Brodo (1976) (Fig. 180a–b)

COMMON NAME: Star-tipped reindeer lichen

STRIKING FIELD CHARACTERS: **Thallus** taking the form of a compact, subglobose head that is densely branched with no apparent main branches. Branches arising from the same axis are unequal in length.

Figure 180a–b. *Cladonia stellaris* (Opiz) Brodo (1976)

MACRO AND MICRO DESCRIPTION: K-, KC+ yellow, P-, UV+.

HABITAT AND ROLE: *Cladonia stellaris* grows on soil and humus, mostly in clumps at the edges of spruce forests at tree line. It commonly forms extensive mats in open woodlands, where the spruce trees are quite stunted in their own growth.

FAMILY: Cladoniaceae

GENUS AND SPECIES: *Cladonia bellidiflora* (Ach.) Schaer. (1823) (Fig. 181a–b)

COMMON NAME: False "British soldiers"

STRIKING FIELD CHARACTERS: **Thallus** greenish yellow and with stalklike projections (podetia) that are covered with tiny, leaf-shaped structures (squamules) and have bright red to scarlet caps (the fungal fruiting bodies). The podetia are 1–3 cm tall and have yellow bases.

MACRO AND MICRO DESCRIPTION: K-, KC+ yellow, P-, UV+/-.

HABITAT AND ROLE: On soil or humus, this beautiful lichen is seen fairly commonly in the spruce forests of the Alaska interior. There are many species of red-fruited *Cladonia*,

Figure 181a–b. *Cladonia bellidiflora* (Ach.) Schaer. (1823)

but none of these other species is as common or has as densely squamulose podetia (which usually end in a narrow cup) as *C. bellidiflora.*

FAMILY: Cladoniaceae

GENUS AND SPECIES: *Cladonia deformis* (L.) Hoffm. (1796) (Fig. 182)

COMMON NAME: Golf tee lichen or red-fruited trumpet lichen

STRIKING FIELD CHARACTERS: **Thallus** grayish green to yellow-green, with projecting stalks (podetia) resembling miniature golf tees or trumpets. The terminal "cups" of these "tees" are rimmed with brownish red marginal proliferations.

Figure 182. *Cladonia deformis* (L.) Hoffm. (1796)

MACRO AND MICRO DESCRIPTION: K-, KC+ yellow, P-.

HABITAT AND ROLE: *Cladonia deformis* typically occurs on soil, rich humus, and rotting wood. *C. sulphurina* (Michaux) Fr. has similar habitat requirements and is rather similar in appearance, but the two species are not often found together. *C. sulphurina* has larger basal squamules and *C. deformis* stouter, more irregularly shaped cups and more prominent fissures, splits, and marginal proliferations on the thallus. *C. borealis* S. Stenroos, another common red-fruited lichen, also occurs in habitats similar to those of *C. deformis*. However, *C. borealis* has shorter, wider trumpets (looking more like goblets) with scarlet margins.

FAMILY: Parmeliaceae

GENUS AND SPECIES: *Dactylina arctica* (Hook. f.) Nyl. (1860) (Fig. 183a–b)

COMMON NAME: Arctic fingers

STRIKING FIELD CHARACTERS: **Thallus** usually dull yellow-green, but becoming brown in exposed situations, consisting of inflated, hollow, fingerlike, upward-projecting elements, these with a smooth surface and often more than 2 cm tall, infrequently branched.

MACRO & MICRO DESCRIPTION: K-, C+ (rarely C-), KC+ yellow, P- or P+ red-orange.

HABITAT AND ROLE: On the ground among alpine and arctic alpine mosses and various other lichens; often occurring at higher elevations or on drier sites along with dryads and ericaceous heaths. There are three other species of finger lichens. *D. beringica* C.D. Bird & J.W. Thomson often occurs together with *D. arctica*, and the two must

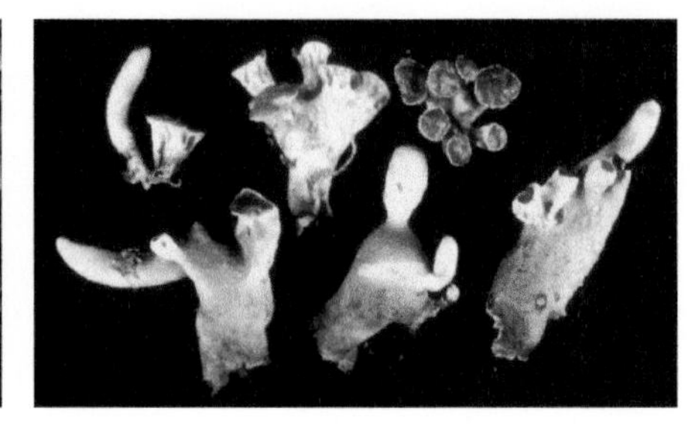

Figure 183a–b. *Dactylina arctica* (Hook. f.) Nyl. (1860)

be distinguished chemically. *D. ramulosa* (Hook.) Tuck. and *D. madreporiformis* (Ach.) Tuck. are well branched and shorter (<2 cm tall). The interior of *D. madreporiformis* is cobwebby, while *D. ramulosa* is hollow, at least in part. *D. ramulosa* has short lateral branches, black dots (pycnidia) present on the surface of the thallus, and a thin frosting of violet-colored crystals (pruina). All of these features are uncommon or completely lacking in *D. madreporiformis*.

FAMILY: Stereocaulaceae

GENUS AND SPECIES: *Stereocaulon tomentosum* Th. Fr. (1825) (Fig. 184a–b)

COMMON NAME: Downy land coral

Figure 184a–b. *Stereocaulon tomentosum* Th. Fr. (1825)

STRIKING FIELD CHARACTERS: **Thallus** silvery gray, branched, the branches rough with abundant, small reddish brown discs (apothecia) present, thickly tomentose.

MACRO AND MICRO DESCRIPTION: K-, C+ (rarely C-), KC+ yellow, P- or P+ red-orange. **Spores** cylindrico-fusiform, spirally twisted, four-celled generally, rarely to eight-celled, 20–35 x 2.5–3 µm.

HABITAT AND ROLE: *Stereocaulon tomentosum* is common on the ground between rocks, forming cushions at the edges of forests, particularly along stream courses, and is often sparsely covered by plants with which it is associated. Many species in this genus form specialized globose to irregularly shaped pink to black structures (cephalodia) that contain nitrogen-fixing cyanobacteria. There are at least 26 species of land coral lichens in Alaska. Some other common species in the Alaska interior include *S. paschale* (L.) Hoffm., *S. alpinum* Laurer ex Funck, *S. glareosum* (Savicz) H. Magn., and *S. grande* (H. Magn.) H. Magn. *S. paschale* is most similar to *S. tomentosum*, but has conspicuous cephalodia, and there are larger, fewer, and flatter discs present on the thallus.

Basidiolichens: Coral

FAMILY: Clavariaceae

GENUS AND SPECIES: *Multiclavula mucida* (Pers.) R.H. Petersen (1967) (Fig. 185a–b)
= *Clavaria mucida* Pers. (1797)
= *Lentaria mucida* (Pers.) Corner (1950)

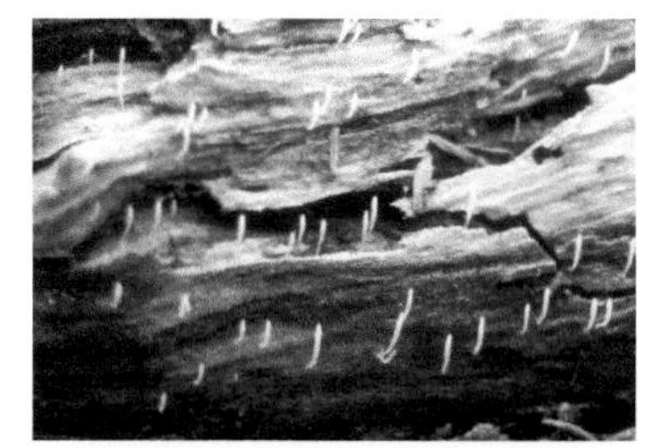

Figure 185a–b. *Multiclavula mucida* (Pers.) R.H. Petersen (1967)

COMMON NAME: Club lichen

STRIKING FIELD CHARACTERS: **Thallus** consisting of single to rarely simply branched, short coral-like clubs, these often occurring in groups.

MACRO AND MICRO DESCRIPTION: **Spores** are oblong-ellipsoid, thin-walled, smooth, nonamyloid, with variously sized vacuoles, 6–7.5 x 2–3 µm.

HABITAT AND ROLE: *Multiclavula mucida* is a lichenized coral fungus. The fruiting bodies of the fungus originate from a thin, bluish green and glistening layer that occurs over the surface of wet disturbed soils (along roadsides) or on old wet, smooth rotting logs in dense forests.

Basidiolichens: Agaric

FAMILY: Tricholomataceae

GENUS AND SPECIES: *Lichenomphalia alpina* (Britzelm.) Redhead, Lutzoni, Moncalvo & Vilgalys (2002) (Fig. 186a–b)
= *Omphalina alpina* (Britz.) Bresinsky and Stangl
= *Omphalina luteovitellinia* (Pil. & Nannf.) M. Lange

COMMON NAME: Yellow basidiolichen or yellow mushroom lichen

Figure 186a–b. *Lichenomphalia alpina* (Britzelm.) Redhead, Lutzoni, Moncalvo and Vilgalys (2002)

STRIKING FIELD CHARACTERS: **Thallus** taking the form of small, bright yellow cap and stipe, single to scattered mushroomlike structures, cap and stalk concolorous.

MACRO AND MICRO DESCRIPTION: **Spores** 7.5–10 x 4.5–5.5 µm.

HABITAT AND ROLE: The brightly colored fruiting body of this basidiolichen stands out in sharp contrast to the dull gray-green, blue-green to deep olive-green background of the humus-rich (peaty) soil upon which it occurs. *Lichenomphalia alpina* is most apt to be seen at higher elevations (alpine zones) in and around peaty soils at the base of granite rock outcrops.

FAMILY: Tricholomataceae

GENUS AND SPECIES: *Lichenomphalia hudsoniana* (H.S. Jenn.) Redhead, Lutzoni, Moncalvo & Vilgalys (2002) (Fig. 187a–b)

= *Omphalia hudsoniana* (Jen.) Big.

= *Omphalina hudsoniana* (Jen.) Big.

COMMON NAME: Hudson's basidiolichen

Figure 187a–b. *Lichenomphalia hudsoniana* (H.S. Jenn.) Redhead, Lutzoni, Moncalvo and Vilgalys (2002)

STRIKING FIELD CHARACTERS: **Thallus** taking the form of a mushroomlike structure, on mossy to rotting woody substrates; with a white, cream, to dull butter yellow cap and a whitish, buff light tan to light olive stalk; and always fruiting from a gray-green foliose to squamulose lichen thallus.

MACRO AND MICRO DESCRIPTION: **Spores** 8.5–10 X 4.5–5.2 µm.

HABITAT AND ROLE: *Lichenomphalia hudsoniana* is another peculiar lichen with a mushroomlike fruiting body. This species always grows on rich humus or mossy or rotting woody (delignified) logs and stumps in partly to completely shaded sites. It is a circumpolar species that dips down the western Rocky and Cascade Mountain ranges and eastern mountains.

FAMILY: Tricholomataceae

GENUS AND SPECIES: *Lichenomphalia umbellifera* (L.) Redhead, Lutzoni, Moncalvo & Vilgalys (2002) (Fig. 188)

= *Omphalia umbellifera* (L.) Quél.
= *Omphalia ericetorum* (Bull.) S. Lundell (1949)
= *Omphalina ericetorum* (Bull.) M. Lange (1955)
= *Omphalina umbellifera* (L.) Quél. (1886)

COMMON NAME: Umbrella basidiolichen

Figure 188. *Lichenomphalia umbellifera* (L.) Redhead, Lutzoni, Moncalvo and Vilgalys (2002)

STRIKING FIELD CHARACTERS: **Thallus** taking the form of a mushroomlike structure. The often deeply cupped (infundibuliform) cap is buff, light tan to light olive to white (in age), flattened (plane) to depressed. The widely spaced and decurrent gills appear to run down onto the concolorous stalk, which is generally darker at its apex. The lichenized portion consists of very small, subglobose to egg-shaped (ellipsoid) structures that form a blue-green "beaded" and wet-looking layer at the base of the mushroomlike fruiting body.

MACRO AND MICRO DESCRIPTION: **Spores** 7–8 x 6–7 µm.

HABITAT AND ROLE: *Lichenomphalia umbellifera* is widely distributed in the Alaska interior, where it typically occurs on old, decaying conifer logs (often with other lichens present) or on humic moss-covered soil.

Plasmodial Slime Molds (Mycetozoans)

In addition to the Deuteromycota, Chytridiomycota, Glomeromycota, Zygomycota, Ascomycota, and Basidiomycota, there is one other fungi-like organismal group to be encountered in the Alaska interior—the slime molds (see Fig. 16q). Although these organisms do not have an attractive name, many slime molds (or myxomycetes, as they are known to mycologists) produce fruiting bodies that exhibit incredibly diverse forms and colors and are sometimes objects of considerable beauty. Most slime mold fruiting bodies are quite small, often no more than 1 or 2 mm in height. Fruiting bodies take the form of tiny goblets, globes, plumes, or other shapes more difficult to characterize. Some typically occur in tightly packed clusters, while others are scattered or solitary. So you just have to get down on all fours and look carefully for these ornate little organisms. However, several "gigantic" species have large sporocarps up to 10 cm and even more in width. Slime molds have long intrigued and perplexed biologists because they possess characteristics of both fungi and animals. The fruiting bodies and spores they produce resemble those of fungi, but some of their other characteristics, including the capability for locomotion during the feeding (assimilative) stages (myxamaoeba and plasmodium), are normally associated only with animals, particularly single-celled heterotrophic protozoans that move using pseudopodia (false feet), such as the amoeba does. For most of its life, a slime mold exists as a thin, free-living mass of protoplasm, actually a single cell that may cover several square centimeters and, as the name suggests, is viscous and/or slimy to the touch. The mass of protoplasm, called a **plasmodium** (pl. plasmodia) can change form and creep slowly about. As it moves along, the plasmodium feeds upon bacteria and bits of organic matter. Slime mold plasmodia occur in moist, shady places such as within crevices of decaying wood and beneath the partially decayed bark of logs and stumps.

After a period of feeding and growth, the plasmodium of a slime mold moves out of its normal habitat and into a drier, more exposed location. Here it gives rise to one or more fruiting bodies. A slime mold fruiting body typically consists of an outer covering (called a **peridium** if thin or a **cortex** if thick) that encloses a mass of spores. In many species of slime molds, a system of threadlike elements (the **capillitium** or **pseudocapillitium**) is found within the spore mass. The fruiting body may be sessile or stalked, depending upon the particular species of slime mold. In one group of slime molds, deposits of lime (calcium carbonate) are found on or within the capillitium, peridium, or stalk of the fruiting body. All of these fruiting body features are used in the identification of various species of slime molds.

FAMILY: Trichiaceae

GENUS AND SPECIES: *Arcyria incarnata* (Pers. ex J.F.Gmel.) Pers. (1796) (Fig. 189)

COMMON NAME: Carnival candy slime mold

STRIKING FIELD CHARACTERS: The clusters of elongated, somewhat drooping, bright red fruiting bodies make this slime mold easy to identify. The capillitium is barely attached to the stalk and the entire mass usually blows away soon after the fruiting bodies become mature.

Figure 189. *Arcyria incarnata* (Pers. ex J.F. Gmel.) Pers. (1796)

MACRO AND MICRO DESCRIPTION: Fruiting bodies stalked, clustered, or crowded; cylindrical; bright red at first but becoming dingy (dirty brown); 0.5–1 mm in diameter and 1.5–5 mm tall. **Peridium** quickly disappears, leaving a shallow cup at the base of the fruiting body. **Stalk** short, pink or pale red, giving rise to a shallow cup at the apex. **Capillitium** consisting of a network of threads, these more or less uniform in diameter and marked with cogs, half rings, and transverse bars; bright red at first but fading to brown. **Spores** red to reddish brown in mass, essentially colorless by transmitted light, with a few scattered warts, 6–8 μm in diameter.

HABITAT AND ROLE: Decaying logs and fallen branches.

EDIBILITY, TASTE, AND ODOR: Despite the common name of carnival candy, *Arcyria incarnata* is not edible.

FAMILY: Physaraceae

GENUS AND SPECIES: *Badhamia utricularis* (Bull.) Berk. (1852) (Fig. 190a–b

COMMON NAME: Fungus-loving slime mold

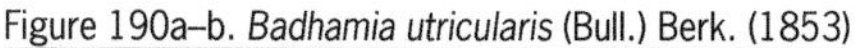

Figure 190a–b. *Badhamia utricularis* (Bull.) Berk. (1853)

STRIKING FIELD CHARACTERS: The relatively large, iridescent, globose, clustered or heaped sporangia that are almost invariably found in association with old fruiting bodies of wood-decaying fungi make this an easy slime mold to recognize.

MACRO AND MICRO DESCRIPTION: **Fruiting bodies** stalked or (more rarely) sessile; clustered or heaped; globose or ovate; blue-gray, iridescent violet or cinereous (ash-gray); 0.5–1 mm in diameter. **Peridium** membranous, iridescent, smooth or somewhat roughened. **Stalks** beige, yellow, or reddish brown; weak, often prostrate. **Capillitium** consisting of a network of calcareous tubes of a more or less uniform diameter, white. **Spores** dull blackish brown in mass, bright violet-brown by transmitted light, warted, 10–14 µm in diameter.

HABITAT AND ROLE: Bark of decaying logs; often associated with the old fruiting bodies of wood-decaying fungi. This slime mold actually feeds upon the fruiting bodies of fleshy fungi and can consume an entire mushroom in a few hours.

EDIBILITY, TASTE, AND ODOR: Although not known to be poisonous, very few slime molds have been consumed as human food. *Badhamia utricularis* does not look like something one would want to eat!

FAMILY: Ceratiomyxaceae

GENUS AND SPECIES: *Ceratiomyxa fruticulosa* (O.F. Müll.) T. Macbr. (1899) (Fig. 191)

COMMON NAME: Coral slime mold

STRIKING FIELD CHARACTERS: The fruiting bodies of *Ceratiomyxa fruticulosa* are unlike those of any other slime mold and somewhat resemble those produced by the coral fungi. When still immature, the fruiting bodies are transparent watery to translucent.

MACRO AND MICRO DESCRIPTION: **Fruiting bodies** consisting of a series of erect, fingerlike columns, which are simple or more commonly branched but sometimes taking the form of a poroid or effused structure, translucent, white or pale yellow and creamy. **Peridium** lacking. **Stalk** lacking. **Capillitium** lacking. **Spores** white in mass, colorless by transmitted light, smooth, round to elliptical, with each spore borne on an individual stalk, 10–13 x 6–7 µm.

Figure 191. *Ceratiomyxa fruticulosa* (O.F.Müll.) T. Macbr. (1799)

HABITAT AND ROLE: On decaying wood or more rarely dead bark or leaf litter. After a period of rainy weather, this species is often exceedingly common on woody debris in forests throughout Alaska.

EDIBILITY, TASTE, AND ODOR: Not edible.

FAMILY: Physaraceae

GENUS AND SPECIES: *Craterium leucocephalum* (Pers.) Ditmar (1813) (Fig. 192a-d)

COMMON NAME: Stalked cup slime mold

STRIKING FIELD CHARACTERS: Old fruiting bodies of *Craterium leucocephalum* are goblet-shaped. Because they are quite small, this feature is best observed with a lens or magnifying glass.

Figure 192a–d. *Craterium leucocephalum* (Pers. ex J.F. Gmel.) *Ditmar* (1812)

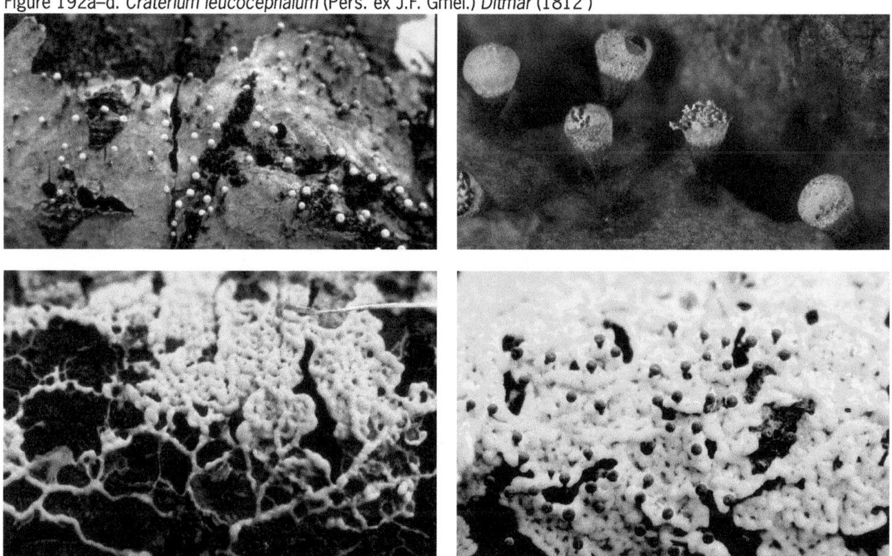

MACRO AND MICRO DESCRIPTION: **Fruiting bodies** stalked, the upper portion globose or obovate to cylindrical, 0.3–0.7 mm in diameter and 1–1.5 mm tall. **Peridium** white and fragile above, ochraceous to yellow-brown or reddish brown below. **Capillitium** consisting of large, irregular white or ochraceous lime nodes connected by slender colorless filaments. **Spores** black in mass, violaceous brown by transmitted light, minutely spiny, 8–9 µm in diameter.

HABITAT AND ROLE: Found on dead leaves and twigs on the forest floor.

EDIBILITY, TASTE, AND ODOR: Like virtually all other slime molds, *Craterium leucocephalum* is much too small to be considered for human consumption.

FAMILY: Cribrariaceae

GENUS AND SPECIES: *Cribraria argillacea* (Pers. ex J.F.Gmel.) Pers. (1794) (Fig. 193)

COMMON NAME: Lead shot slime mold

STRIKING FIELD CHARACTERS: The dense clusters of globose sporangia, which are lead-colored when immature, are distinctive. Fruitings of this species tend to be large and may extend over half a meter along a decaying log.

Figure 193. *Cribraria argillacea* (Pers. ex J.F.Gmel.) Pers. (1794)

MACRO AND MICRO DESCRIPTION: Fruiting bodies short-stalked to almost sessile, clustered and often occurring in large fruitings, globose, shining lead-colored at first and then becoming dull clay-colored, 0.5–1 mm in diameter. **Peridium** quickly disappearing above except for a thin, netlike covering over the spore mass but persisting below as a deep, cuplike structure, often somewhat iridescent throughout. **Stalk** short, usually no more than 1 mm long, dark brown to black, furrowed. **Capillitium** lacking. **Spores** clay-colored in mass, very pale by transmitted light, nearly smooth, 6–8 μm in diameter.

HABITAT AND ROLE: Decaying conifer logs and stumps.

EDIBILITY, TASTE, AND ODOR: Not edible, although sometimes consumed by small invertebrates.

FAMILY: Physaraceae

GENUS AND SPECIES: *Fuligo septica* (L.) F.H. Wigg. (1780) (Fig. 194)

COMMON NAME: Scrambled-egg slime mold

Figure 194. *Fuligo septica* (L.) F.H. Wigg. (1780)

STRIKING FIELD CHARACTERS: Because of its large size and often bright yellow color, *Fuligo septica* is one of the most conspicuous of all slime molds.

MACRO AND MICRO DESCRIPTION: Fruiting bodies cushion-shaped, white to pale or bright pink or red to bright yellow, usually quite large, 2–20 cm in extent and 1–3 cm thick. Outer portion (**cortex**) of fruiting body is usually fairly thick, calcareous, and fragile. **Capillitium** consisting of a system of white, yellow, or reddish yellow-lime nodes connected by colorless filaments. **Spores** dark gray or dull black in mass, light purplish brown by transmitted light, nearly smooth to minutely spiny, 6–9 µm in diameter.

HABITAT AND ROLE: Found on decaying wood and bark, forest floor litter, wood debris, and soil. The fruiting bodies are not uncommon on old sawdust and the beds of mulch found around ornamental shrubs.

EDIBILITY, TASTE, AND ODOR: *Fuligo septica* is one of the few slime molds that has been reported to have been consumed by humans. However, it is the stage at which the plasmodium is transforming into the fruiting body and not the mature fruiting body that is eaten.

FAMILY: Lycogalaceae

GENUS AND SPECIES: *Lycogala epidendrum* (L.) Fr. (1829) (Fig. 195a-c)

Figure 195a–c. *Lycogala epidendrum* (L.) Fr. (1829)

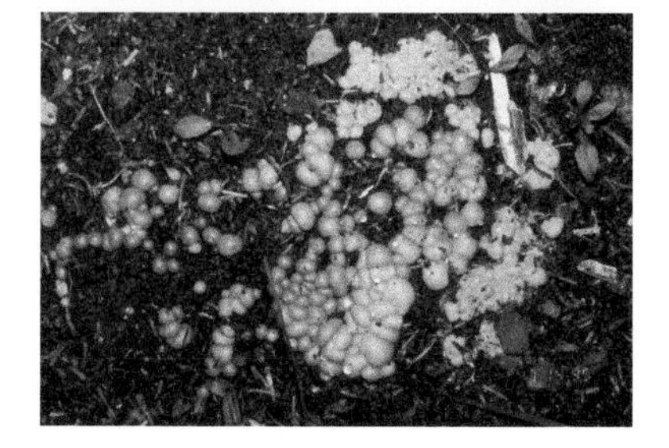

COMMON NAME: Wolf's milk slime mold

STRIKING FIELD CHARACTERS: The fruiting bodies resemble small puffballs and are unlikely to be confused with those of any other slime mold.

MACRO AND MICRO DESCRIPTION: Fruiting bodies are subglobose to depressed-spherical or irregular from pressure, pinkish gray or yellowish brown to deep olivaceous or nearly black, 3–15 mm broad. The outer covering (**cortex**) is warted or merely roughened, relatively thin and fragile, especially above. **Pseudocapillitium** consisting of a network of long, branching, and anastomosing flattened tubes marked with conspicuous folds and wrinkles. **Spores,** at first pink or red in mass, changing to pale ochraceous or pallid, colorless by transmitted light, reticulate, 6–8 µm in diameter.

HABITAT AND ROLE: Scattered or clustered on decaying wood and less commonly on bark of logs and stumps. This is one of the most widely distributed and best-known of all slime molds.

EDIBILITY, TASTE, AND ODOR: Although not consumed by humans, old fruiting bodies of *Lycogala epidendrum* often contain the larvae and/or adults of small beetles. These slime mold beetles feed upon the spore mass and also use the fruiting body as a breeding site.

FAMILY: Didymiaceae

GENUS AND SPECIES: *Mucilago crustacea* F.H. Wigg. (1780) (Fig. 196a–b)

Figure 196a–b. *Mucilago crustacea* F.H. Wigg. (1780)

COMMON NAME: White crust slime mold

STRIKING FIELD CHARACTERS: No other large white slime mold is likely to be found fruiting upon living plants, the usual habitat for *Mucilago crustacea.*

MACRO AND MICRO DESCRIPTION: **Fruiting bodies** usually solitary, cushion-shaped, consisting of numerous anastomosing tubes with membranous walls, pure white to creamy white or pale ochraceous, 1–7 cm long, 1–5 cm wide, and 1–2 cm thick. Outer portion (**cortex**) of fruiting body is dense, spongy, or flaky and composed of crystals of calcium carbonate. **Capillitium** consisting of a network of dark, branching, and anastomosing filaments, often with expanded portions. **Spores** black in mass, blackish brown or occasionally bright purplish brown by transmitted light, densely and unevenly warted or spiny, 11–13 µm in diameter.

HABITAT AND ROLE: Found on low-growing living plants, forest floor litter, and less commonly on decaying wood. This slime mold is sometimes surprisingly common in areas of tundra.

EDIBILITY, TASTE, AND ODOR: Although it is sometimes large enough to be considered as an item for the table, there are apparently no reports of *Mucilago crustacea* ever being collected and consumed by humans.

FAMILY: Trichiaceae

Figure 197. *Perichaena minor* (G. Lister) Hagelst. (1943)

GENUS AND SPECIES: *Perichaena minor* (G. Lister) Hagelst. (1943) (Fig. 197)

COMMON NAME: Warted slime mold

STRIKING FIELD CHARACTERS: The fruiting bodies of this slime mold are relatively small and thus easily overlooked. However, the large, dark, wartlike structures on the peridium are distinctive.

MACRO AND MICRO DESCRIPTION: **Fruiting bodies** stalked or (very rarely) sessile, scattered, the upper portion globose and characterized by the presence of a number of dark brown to black wartlike structures, the entire fruiting body less than 1 mm tall. **Peridium** membranous but adorned with a number of prominent dark, wartlike structures, dull olive-yellow. **Stalk** short, dark brown or black. **Capillitium** consisting of a network of threads of a more or less uniform diameter, with a few faint spirals covered with short spines. **Spores** olive-yellow in mass, pale greenish yellow by transmitted light, faintly warted, 9–11 µm in diameter.

HABITAT AND ROLE: Sometimes occurring on the bark of living trees but more common on the litter from broadleaf trees such as aspen.

EDIBILITY, TASTE, AND ODOR: This mycetozoan is much too small to be considered as food.

FAMILY: Physaraceae

GENUS AND SPECIES: *Physarum bivalve* Pers. (1795) (Fig. 198)

COMMON NAME: Clamshell slime mold

STRIKING FIELD CHARACTERS: The fruiting bodies of this slime mold often have the shape of miniature clamshells, which accounts for the common name. This is not always the case for an entire fruiting, but usually a few examples with the distinctive shape can be found.

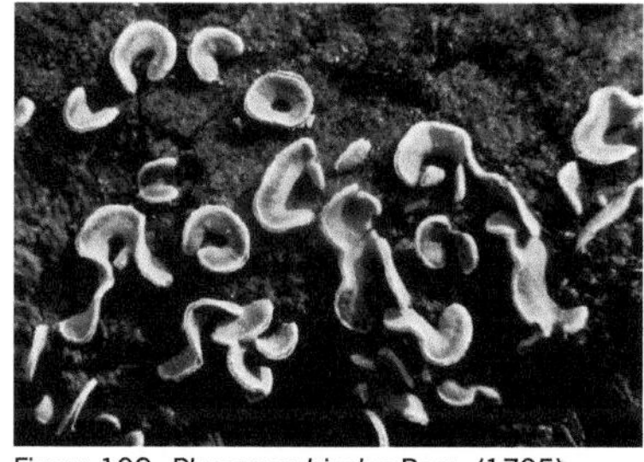

Figure 198. *Physarum bivalve* Pers. (1795)

MACRO AND MICRO DESCRIPTION: **Fruiting bodies** sessile, elongated, strongly compressed laterally, scattered, white, 0.7–1 mm tall and sometimes as much as 20 mm long. **Peridium** consisting of two layers, the inner delicate, membranous, and the outer thicker and encrusted with white lime, opening along the top of the fruiting body by a preformed longitudinal split. **Capillitium** consists of large, rounded lime nodes connected by thin, colorless filaments. **Spores** black in mass, dull violet brown by transmitted light, finely warted, 8–10 µm in diameter.

HABITAT AND ROLE: Usually occurring on leaf litter or more rarely on bryophytes.

EDIBILITY, TASTE, AND ODOR: Not edible.

FAMILY: Trichiaceae

GENUS AND SPECIES: *Trichia varia* (Pers. ex J.F. Gmel.) Pers. (1794) (Fig. 199)

COMMON NAME: Yellow wood drops slime mold

STRIKING FIELD CHARACTERS: Although the fruiting bodies are rather small, they are often easy to spot because their yellow to orange-yellow color causes them to stand out in sharp contrast to the decaying bark or wood upon which they occur.

Figure 199. *Trichia varia* (Pers. ex J.F. Gmel.) Pers. (1794)

MACRO AND MICRO DESCRIPTION: Fruiting bodies usually occurring in small clusters; sessile or occasionally with very short, dark stalks; globose to ovate or sometimes somewhat elongated; yellow to orange yellow or sometimes yellow brown; 0.5–1 mm in diameter. **Peridium** membranous and translucent or somewhat cartilaginous when thickened with granular deposits. **Capillitium** consisting of a tangled mass of threadlike elements bearing irregular spiral bands and with pointed tips. **Spores** yellow to orange-yellow in mass, dull pale yellow by transmitted light, delicately warted, 12–14 μm in diameter.

HABITAT AND ROLE: Occurs on decaying wood and bark of stumps and logs.

EDIBILITY, TASTE, AND ODOR: Although much too small to be considered for human consumption, the fruiting bodies of *Trichia varia* are sometimes consumed by small invertebrates, including mites and beetles.

Bryophytes: The Mosses and Liverworts

Introduction

Mosses, liverworts (or hepatics), and hornworts collectively make up the group of plants known as the Bryophytes. Bryophytes are mostly small plants that generally inhabit moist or humid environments. There are, however, many examples known from more arid environments such as the hot, dry deserts, higher alpine regions, and polar environments. They grow on many substrates such as rock; soil; organic materials including bark, wood, dung, and bone; and on manmade substrates such as old cars, roofs of buildings, old clothing, bricks, and concrete.

Like all terrestrial organisms, bryophytes require free water for survival. However, unlike fungi, they are typical green plants with chlorophyll *a* and *b* as their photosynthetic pigments, starch as an energy-storage product, cellulose cell walls, and a cuticle to help limit surface water loss.

Unlike the flowering plants, conifers, ferns, psilotophytes (the psilotums), and lycopods (the club mosses), all of which have a well-developed internal vascular system for water and nutrient conduction and for providing structural support, bryophytes have a relatively simple internal structure. Water conduction is mostly over the surface of the plants by capillary action.

The life history of bryophytes involves an alternation of gametophyte and sporophyte generations. The leafy, or thallose, gametophyte—usually the bulk of the plant—gives rise to a sporophyte that produces spores as the result of sexual reproduction. Male (antheridia) and female (archegonia) reproductive structures are borne on the gametophyte plant, which may produce either male or female only (i.e., each sex on a separate plant, or both on one plant). After release from the mature antheridium, motile sperm swim in a film of water to the receptive archegonium. Attraction is by chemical means. Fertilization of a mature egg cell in the archegonium by a single sperm results in the development of the sporophyte, which is quite different in appearance to the gametophyte. The sporophyte in nearly all bryophytes consists of a foot (attachment region), a seta (stalk), and a capsule in which the spores develop.

The sporophytes are often quite conspicuous but may be small and hard to see, particularly when the seta is very short. In some mosses, such as the seldom-found *Buxbaumia*, the capsule may be very large, dwarfing the gametophyte plant.

In mosses, the sporophyte is borne either at the tip of the growing shoot (acrocarpous arrangement) or on a lateral branch or some distance from the growing shoot apex (pleurocarpous arrangement). In the liverworts, a variety of sporophyte types is found. In *Riccia*, the sporophyte lacks a seta and the capsule develops within the thallus or plant body. In some thallose forms, such as *Marchantia*, the sporophytes develop on special structures elevated above the body of the plant. In many leafy forms, the sporophyte develops within modified leaves (referred to as the perianth) and only becomes visible for a short time when it is elevated above the perianth at maturity. In some leafy forms the capsule develops in a special underground saclike structure (marsupium). In the hornworts, the sporophyte may be fingerlike and is quite different in many features to other bryophyte capsules.

In addition to sexual reproduction, many bryophytes are able to reproduce asexually through fragmentation of the gametophyte plants or by the production of specialized asexual propagules (gemmae, tubers, bulbils, specialized branchlets) that are borne on the gametophyte plants, on leaves, or the stems, or sometimes as special underground structures (tubers) on the rhizoids (fine rootlike structures that serve primarily to anchor the plants to their substrate). Some thallose liverworts, such as *Marchantia*, have specialized cuplike structures on their upper surface in which asexual gemmae are borne. In some leafy liverworts, specialized gemmae are often found on the margins or tips of the leaves. Asexual reproduction is a very successful mechanism for both dispersal and for the colonization of new and favorable habitats. The trade-off is the lack of genetic diversity brought about through the absence of sexual reproduction.

Differences Between Mosses, Liverworts, and Hornworts

Mosses differ from the liverworts and hornworts in many ways. No mosses are thallose, whereas there are many thallose liverworts, as are all of the hornworts. Leafy liverworts can easily be distinguished from mosses using a hand lens or by eye. More detailed microscopic examination reveals many beautiful and unique features that are often necessary for complete identification.

Mosses sometimes have their leaves arranged in two or three rows along the stems, but the number of rows is usually more. The leaves are never lobed and they often have a costa (midvein). The costa may be absent, single, long or short, or sometimes double and may be difficult to observe. In many species the costa extends well up the leaf, may end in the leaf tip, or may extend beyond the tip as a hair point. The fine rhizoids that attach the plants to the substrate are usually reddish brown to dark brown, and are branched. The gametophyte leaves at the base of the sporophyte may be larger than other stem or branch leaves, but a perianth or sleevelike structure that is made up of fused leaves, as found in many leafy liverworts, is never present in mosses. The seta of the sporophyte is usually pigmented (brown, yellowish, reddish purple), is stiff and wiry, and persists long after the spores are shed from the capsule (sporangium). The capsule (sporangium) develops after the seta elongates and usually opens by an apical lid (operculum). In *Andreaea*, the capsule opens by four longitudinal slits and in the highly unusual *Takakia* (initially thought to be a liverwort, before capsules were found) the capsule opens by a spiral slit. Only spores develop within the capsule of a moss. The mouth of the capsule usually has one or two rows of specialized structures (peristome teeth), which aid in controlling spore dispersal. Sometimes the peristome teeth are entirely absent. During development the capsule, located at the top of the elongating seta, is covered by a cap (calyptra) derived of tissue from the female reproductive structure (archegonium).

Leafy liverworts have the leaves arranged in two or three rows, are often lobed, and lack a costa. The rhizoids are usually colorless, but are sometimes red, and are unbranched, although they may have short fingerlike lobes at the ends at the point of attachment to the substrate. A perianth sheath is usually developed around the female reproductive structures. The seta is usually colorless and fragile, and collapses when the spores are released from the capsule (sporangium). The sporangium matures before the seta elongates, and in addition to the spores usually contains elongate single-celled elaters (which aid spore dispersal), the walls of which are mostly spirally thickened. Mostly, the capsule opens by longitudinal lines. There is no peristome developed and the calyptra remains at the base of the seta as it elongates to elevate the sporangium.

The hornworts form an isolated evolutionary lineage and may bear little direct relationship to the other bryophytes. They resemble thallose hepatics in form of the gametophyte, which is composed of thin-walled cells and is dorsiventrally flattened. The thallus often contains mucilage-filled cavities that are also often filled with clusters of cyanobacteria (blue-green algae). The antheridia develop initially from an epidermal cell but subsequent development is internal. Similarly, the archegonia develop initially from an epidermal cell but then develop internally. At maturity, the neck of the archegonium projects through the thallus surface. The sporophyte is unique, resembling an elongate, tapered cylinder. Its base is surrounded by a collarlike sheath that is derived from the surface of the gametophyte. There is no calyptra. In most species, growth of the sporophyte is indeterminate, growing from a meristem or growing region. There is a large basal foot region with, usually, a region of growth (meristem) positioned above it, and the tissue that gives rise to the spores and pseudo-elaters, simple one-to-several-celled filaments mixed with the spores, positioned above the meristem region. Spores mature from the apex of the sporophyte toward the base. The pseudo-elaters, although similar to the elaters of liverworts, form in a completely different way, in layers alternating with spores. The sporophytes are green and at maturity split lengthways, into two halves that remain joined for some time at the tip, to release the spores. Some hornworts produce asexual propagules on the surface or along the margins of the thallus. Only one hornwort, *Anthoceros punctatus*, has been recorded from Alaska and this group is not further considered here.

Ecological Importance of Bryophytes

Bryophytes are significant components of the biodiversity of many habitats, such as the tundra, boreal forests, peatlands, and wetlands. They are able to colonize hard substrates such as rock surfaces and bark, and are primary colonists of soil surfaces. They play a significant part in the stabilization of substrates that may later become colonized by vascular plants. Many bryophytes have a high tolerance for long periods of desiccation. They provide specific substrates for some fungi, shelter for small mammals and a host of invertebrate animals, a seedbed for higher plants, nest materials for birds and animals, as well as ground-cover insulation. They play a significant part in the regulation of nutrient cycling, absorption of moisture, and in the carbon and nitrogen balance of the ecosystem. They are also aesthetically pleasing and beautiful.

Around 640 mosses and 280 hepatics have been reported for the bryophyte flora of Alaska. Hepatics are generally less conspicuous than the mosses and many are difficult to identify without some specialist experience and access to a good microscope.

A significant proportion of the Alaska bryoflora can be found within the Alaska interior. As with the vascular flora (Hultén 1968), there are many species whose range extension into Alaska is found only in the southeastern and southcentral regions. Similarly, some taxa are confined to the northern and interior regions.

In the Alaska interior, bryophytes are found in a wide variety of habitats: from very wet to strongly arid; alkaline to acid; on bark and wood; on rock and soil; the forest floor; open shrublands and grasslands; tundra; on old dung and other organic substrates. Many are rarely encountered. Others are very common. Some are opportunist colonizers of bare ground, such as along fresh road cuts, land slips, the soil around uprooted trees, and exposed mineral soils following fires. As a group of plants, they are highly adaptive and many species exhibit wide ecological amplitudes.

Some are typically found as colonists of recently exposed soils; e.g., the mosses *Ceratodon purpureus* and *Funaria hygrometrica*, the latter being one of the most obvious early colonists of burnt ground. *Polytrichum* species and a number of lichens follow closely in the colonization succession. Other common early colonists of bare soils, perhaps also to be considered as weedy species, include *Leptobryum*, *Pohlia*, and some *Bryum* species.

Many mosses (e.g., *Andreaea*, *Grimmia*, *Schistidium*, *Hedwigia*) are found primarily on rock and soil over rock surfaces. There can be a significant difference in the flora of calcareous or siliceous rocks and soils. Species growing on rock, or open bare soil surfaces, are subject to extreme microclimatic conditions. Dry rock surface temperatures in full sun on a fine day can reach in excess of 40°C (105°F) and overnight fall to below freezing point (0°C/32°F). Moisture availability through dew, rain, snowfall, or seepage can be critical for survival, growth, and sexual reproduction. Surface chemistry, microtopography, and stability, together with moisture availability, are important for establishment.

A number of mosses and liverworts are found on trees, on bark or wood, and often on the forest floor growing on old bark or wood. Species of *Orthotrichum* occur commonly on the bark of alder, poplar, and aspen. *Pylaisiella* and *Hypnum* are often found forming a distinctive skirt at the base of trees.

A large suite of species is found on the forest floor where they often form extensive mixed carpets. The feather mosses *Pleurozium* and *Hylocomium*, are particularly striking, together with *Dicranum, Tomenthypnum, Ptilium, Polytrichum, Sphagnum*, and many others.

In wetter seepage habitats, *Aulacomnium, Brachythecium, Climacium, Sphagnum, Plagiomnium*, and many other mosses abound. With increasing acidity *Sphagnum* species become dominant. Other striking mosses found in peatlands include, for example, *Tomenthypnum, Dicranum, Helodium, Catascopium, Pohlia*, and *Meesia*.

Tetraplodon and *Splachnum* are found primarily on old organic substrates, particularly rotting dung. The spores are spread by insects that are attracted by a musty chemical aroma to the mature capsules of the moss, where they pick up the spores and then transfer these to the fresh dung, where the spores are deposited. *Splachnum luteum* and the even more rare *S. rubrum* are particularly striking with their yellow (*S. luteum*) and purple-red (*S. rubrum*) parasol-like extensions of the hypophysis (modified base of the capsule, below the spore sac).

Bryophyte Flora of North America

In the preparation of this section, we frequently refer the new *Moss Flora of North America* published in 2007. Of particular use has been the work of Richard Andrus on *Sphagnum*, Wilf Schofield on *Hypnum*, and Robert Ireland on *Dicranum*. These and many other generic treatments were made available on the website (http://www.mobot.org/plantscience/bfna/bfnamenu.htm) as they were completed. We are grateful to the editor, Richard Zander (Missouri Botanical Garden), for making these prepublication treatments available on the Internet.

Descriptions of Common Moss Species

FAMILY: Thuidiaceae

GENUS AND SPECIES: *Abietinella abietina* (Hedw.) Fleisch. (Fig. 200)

COMMON NAME: Abietinella moss

Figure 200. *Abietinella abietina* (Hedw.) Fleisch.

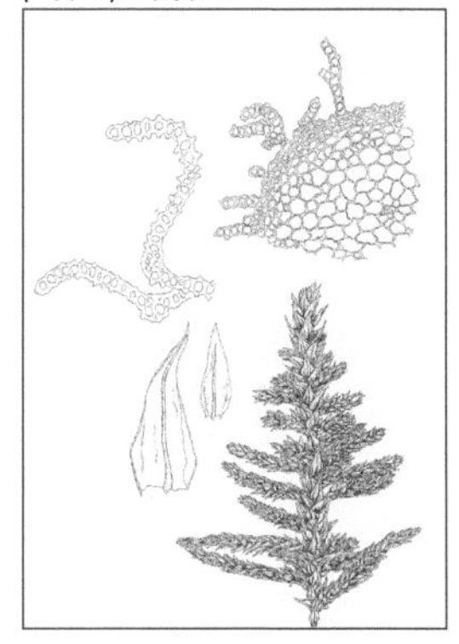

STRIKING FIELD CHARACTERS: *Abietinella* (sometimes placed in *Thuidium*) is primarily a species of well-drained habitats. The rather stiff stems and the usually yellowish brown color, together with the pinnate shoots, are key identifying features.

MACRO AND MICRO DESCRIPTION: Plants forming loose, stiff, green to yellowish or brownish mats. Stems variable in length, 5–10 cm or more long, prostrate to erect-ascending, pinnate; branches 5–12 mm long, terete, tapered towards the tips. Stem paraphyllia abundant, simple or branched, the cells strongly unipapillose. Stem leaves erect-spreading, to 2 mm long, broadly ovate-lanceolate, acuminate, pleated; costa extending to about three-fourths length of leaf; lamina cells strongly unipapillose on each surface. Branch leaves smaller, less than 1 mm long, similar in shape to stem leaves, acute.

Setae to 2.5 cm long, reddish brown, flexuose; capsules to 3 mm long, cylindric, strongly curved, inclined; Sporophytes rare.

HABITAT AND ROLE: Occurs on soil, rock, or peat, mostly in dry habitats. Usually found on sandy soils in calcareous habitats.

FAMILY: Aulacomniaceae

GENUS AND SPECIES: *Aulacomnium palustre* (Hedw.) Schwägr. (Fig. 201)

COMMON NAME: Aulacomnium moss

STRIKING FIELD CHARACTERS: *Aulacomnium palustre* is a common wetland, heathland, and forest species. The yellowish green leaves and stems with numerous brown axillary rhizoids are useful field characteristics. The bright white costa is a striking feature.

MACRO AND MICRO DESCRIPTION: Plants in loose to dense, yellowish green to brownish green tufts. Stems up to 10 cm long, with numerous brown rhizoids (tomentum) arising from the leaf axils, except near the shoot apices. Leaves erect-spreading when moist, somewhat contorted when dry, to 4 mm long, keeled, oblong-lanceolate, acute or acuminate; basal angles of the leaves slightly decurrent (extending down the stem); leaf margins recurved, entire, or serrulate towards the apex; costa single, ending shortly below the apex; lamina cells thick-walled, singly papillose on both surfaces. Setae to 4 cm long, brown; capsules 2–4 mm long, inclined to horizontal, curved.

Figure 201. *Aulacomnium palustre* (Hedw.) Schwaegr.

HABITAT AND ROLE: The plants are often found without capsules, but these sterile stems often bear clusters of small, dark green leaflike asexual propagules at the tip and scattered along the sides of an elongated stalklike extension (pseudopodium) of the stem. These propagules are easily dislodged. *Aulacomnium androgynum* (Hedw.) Schwägr. is a shorter but closely similar and far less common species. It is found primarily on decaying wood and sometimes on soil or soil over rock in woodlands. The small asexual propagules are spindle-shaped and borne in a tight, globular cluster at the tip of the otherwise naked pseudopodium.

FAMILY: Aulacomniaceae

GENUS AND SPECIES: *Aulacomnium turgidum* (Wahlenb.) Schwaegr. (Fig. 202)

COMMON NAME: Turgid aulacomnium moss

Figure 202. *Aulacomnium turgidum* (Wahlenb.) Schwaegr.

STRIKING FIELD CHARACTERS: The turgid, large erect shoots with relatively blunt leaves that are somewhat contorted when dry are conspicuous differentiating features. It is usually found in well-drained sites.

MACRO AND MICRO DESCRIPTION: Plants in yellow-green tufts, stems to 10 cm long, seldom branched. Leaves are imbricate, to 3.5 mm long, 1 mm broad, elongate-ovoid to oblong, concave, and often somewhat channeled at the rounded apex.

HABITAT AND ROLE: It is more a species of well-drained sites at higher elevations and latitudes. It is sometimes found together with the much more abundant and widespread *Aulacomnium palustre*.

FAMILY: Brachytheciaceae

GENUS AND SPECIES: *Brachythecium albicans* (Hedw.) Bruch, Schimp., & W. Gümbel (Fig. 203)

COMMON NAME: Brachythecium moss

Figure 203. *Brachythecium albicans* (Hedw.) Schimp.

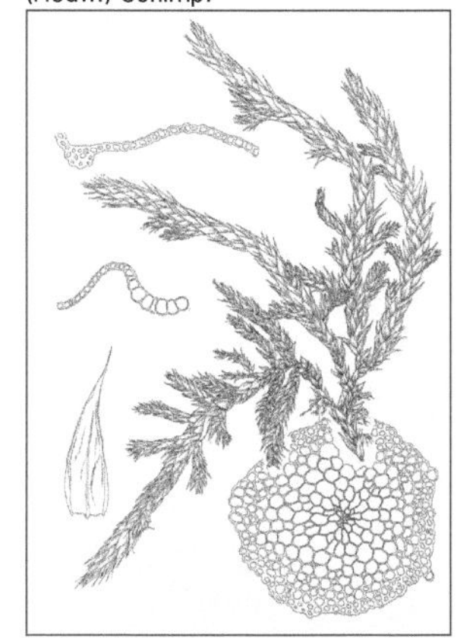

STRIKING FIELD CHARACTERS: *Brachythecium* species are not always easy to differentiate. The leaves are usually glossy, and with *B. albicans* the whitish green coloration is a useful spotting characteristic.

MACRO AND MICRO DESCRIPTION: Plants forming loose mats, light green to whitish green. Stems prostrate, irregularly branched, the branches terete and sometimes julaceous. Leaves imbricate, concave, ovate-lanceolate, tapering to a long slender point, the basal margins decurrent (extending down the stems); costa single, reaching to beyond midleaf; leaf margins entire or nearly so; stem leaves 2–3.5 mm long, about 1 mm broad. Setae to 2 cm long, smooth, reddish brown; capsules inclined to horizontal, curved, about 2 mm long. Sporophytes are rarely seen.

HABITAT AND ROLE: Occurs on rock, soil, litter, and peat, often in grassy habitats.

FAMILY: Brachytheciaceae

GENUS AND SPECIES: *Brachythecium plumosum* (Hedw.) Schimp. (Fig. 204)

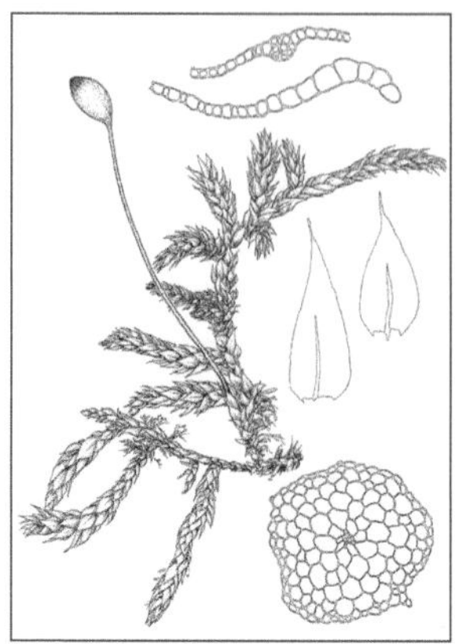

Figure 204. *Brachythecium plumosum* (Hedw.) Schimp.

COMMON NAME: Brachythecium moss

STRIKING FIELD CHARACTERS: *Brachythecium* species are usually quite glossy and are common components of the bryoflora of wet habitats. Distinguishing among the many species is, however, often difficult.

MACRO AND MICRO DESCRIPTION: Plants robust, green to yellowish green or brownish green, glossy. Stems prostrate, forming mats, irregularly to subpinnately branched. Stem leaves imbricate when dry, usually curved to one side of the stem when moist, to 2 mm long, concave, lanceolate to ovate-lanceolate, acuminate, weakly striolate; costa single, reaching about two-thirds length of the leaf; branch leaves scarcely different. Setae to about 1.5 cm long, dark red, roughened by papillae towards the tip or sometimes smooth throughout. Capsules up to 2 mm long, oblong-ovoid, asymmetric, inclined to horizontal, dark brown. Sporophytes are frequent.

HABITAT AND ROLE: This is a plant of wetter habitats, usually on rock. The yellowish green to brownish plants and the more or less smooth, concave leaves that show scarcely any differentiation between stem and branches, are useful characteristics for recognition. The branch tips are often slightly curved to one side.

FAMILY: Bryaceae

GENUS AND SPECIES: *Bryum arcticum* (R. Br.) Bruch & Schimp. (Fig. 205)

Figure 205. *Bryum arcticum* (R.Br.) Bruch & Schimp.

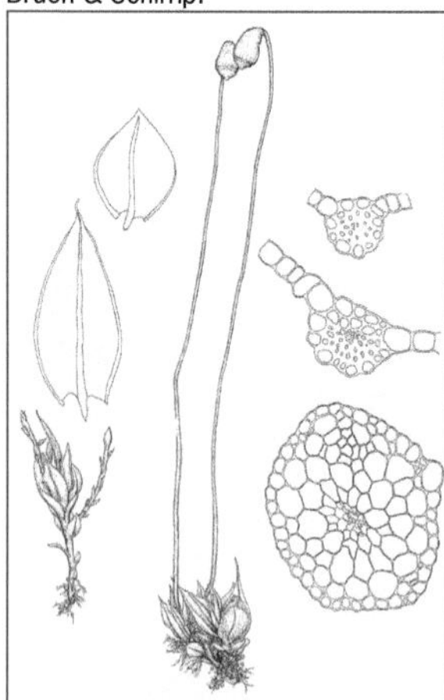

COMMON NAME: Arctic bryum moss

STRIKING FIELD CHARACTERS: Other species of *Bryum* found within the Alaska interior are larger. They may be easily confused at first glance with species of *Pohlia*. Leaves of *Bryum* are typically bordered by narrow elongate cells, but sometimes the border is often indistinct. In *Pohlia*, there is no border and the leaf cells are narrower and more elongate.

MACRO AND MICRO DESCRIPTION: Plants forming yellowish green to reddish green tufts with age, matted together below with brown rhizoids. Stems red, occasionally branched, rather variable in length. Leaves erect-spreading when moist, appressed flexuose when dry, to 2.5 mm long, often shorter; margins bordered by two to three rows of narrow, elongate cells sharply different to the shortly rectangular to elongate-rhomboid cells of the leaf lamina; costa single, reddish, shortly excurrent. Setae to about 3 cm in length, dark reddish brown; capsules yellow-brown, drooping to pendulous, pyriform, slightly asymmetric, the urn about 1.5–2.5 mm long and narrowed at the mouth.

HABITAT AND ROLE: The genus *Bryum* is widespread throughout the world and the determination of species is often exceedingly difficult, or almost impossible, without mature capsules. *B. arcticum* is an arctic-alpine species of soil and rock crevices. *B. argenteum* is one of the small silver-green *Bryum* species that is a common colonist of bare soils, cracks in footpaths, etc. The stems are short, to about 1 cm high, branched, and the plants form dense tufts and mats. It is not uncommonly found fruiting. There are several short-stemmed, green (rather than silver-green) *Bryum* species that commonly show vegetative reproductive propagules in the axils of the stem leaves.

FAMILY: Campyliaceae

GENUS AND SPECIES: *Campylium stellatum* (Hedw.) C. Jens. (Fig. 206)

COMMON NAME: Star campylium moss

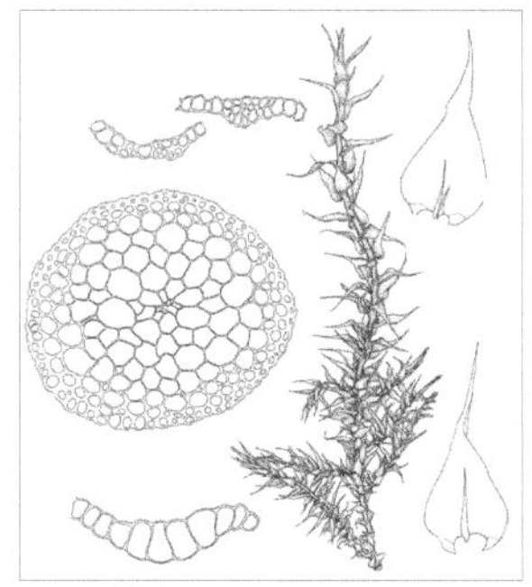

Figure 206. *Campylium stellatum* (Hedw.) C. Jens.

STRIKING FIELD CHARACTERS: The specific name refers to the wide, spreading to squarrose leaves that give the stems a starlike appearance when viewed from above.

MACRO AND MICRO DESCRIPTION: Plants in loose to dense, yellowish green to golden mats, becoming brownish with age. Stems irregularly branched, ascending. Leaves close together to somewhat distant, erect-spreading to squarrose, from a cordate-ovate base gradually narrowed to a long and somewhat channeled point; margins plane, entire or weakly serrulate towards the base; costa absent, single or double, short or reaching to about one-half leaf length; leaf lamina cells narrow elongate, those at the basal angles enlarged, rectangular to rounded, hyaline, clearly differentiated from adjacent lamina cells. Setae to about 3 cm long; capsules 2–3 mm long, curved, narrowed below the mouth when dry.

HABITAT AND ROLE: This is a common moss of moist to boggy areas, particularly in calcareous habitats.

FAMILY: Catoscopiaceae

GENUS AND SPECIES: *Catoscopium nigritum* Brid. (Fig. 207a–b)

COMMON NAME: Black golf-club moss

STRIKING FIELD CHARACTERS: When fruiting, the small green capsules that turn brown and then become glossy black are highly characteristic. The capsules are inclined

Figure 207a–b. *Catoscopium nigritum* (Hedw.) Brid.

at a right angle to the seta, and give the impression of being a small, upturned golf club. The small, narrowly triangular leaves and the dense, dark green to yellowish green tufts are useful identifying features.

MACRO AND MICRO DESCRIPTION: Plants forming dense bright green to brownish green tufts or cushions. Stems erect, very variable in length, to 10 cm or more, sparsely branched, often with numerous rhizoids along at least the lower parts of the stems. Leaves are erect-spreading when moist, 1–2 mm long, lanceolate to narrowly triangular, keeled; costa strong, ending shortly below the apex or at the apex. Setae to about 15 mm long, brown; capsules small, about 1 mm long, subspherical to rather flattened on the lower and rounded on the upper side, with a small mouth, inclined at about 90 degrees, becoming glossy dark brown to black at maturity.

HABITAT AND ROLE: It is a species of moist calcareous habitats.

FAMILY: Ditrichaceae

GENUS AND SPECIES: *Ceratodon purpureus* (Hedw.) Brid. (Fig. 208a–b)

COMMON NAME: Ceratodon moss

STRIKING FIELD CHARACTERS: Occasionally, forms are found where the leaves are almost ovate with a rounded obtuse apex. Quite unlike the typical forms, these aberrant forms are readily identified by a combination of characters, such as the revolute margins; lamina cells being thick-walled and rather rounded-quadrate or short rectangular; the costa, which is prominent on the abaxial surface of the leaf; and the anatomy of the costa. The upper cells of the leaf are often somewhat elongate and the upper margins toothed.

MACRO AND MICRO DESCRIPTION: Plants form dense tufts to extensive mats or carpets. Stems 0.5–2.5 cm high, green to brownish green, yellowish brown, or reddish. Leaves to 2 mm long, lanceolate to ovate-lanceolate, acute to shortly acuminate, keeled; margins strongly recurved or reflexed almost to the apex, irregularly serrulate near the apex; costa strong, prominent on the abaxial surface, ceasing shortly below the apex to excurrent. Setae 1–3 cm long, purple-red or occasionally yellowish orange; capsules

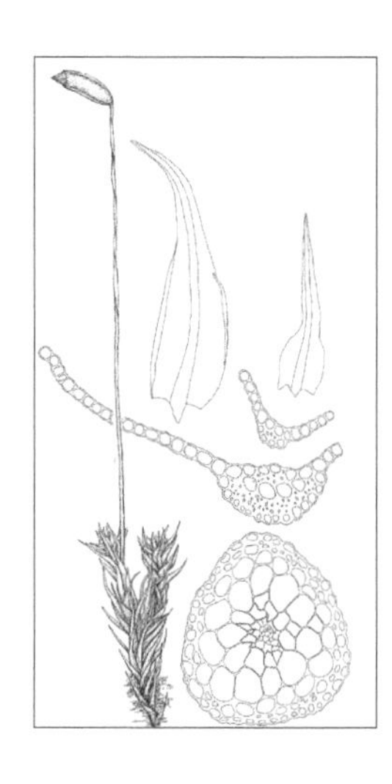

Figure 208a–b. *Ceratodon purpureus* (Hedw.) Brid.

dark red-brown, becoming paler when old, 1.5–2.5 mm long, ovoid-cylindric, the neck constricted and appearing strumose when dry, inclined to horizontal, asymmetric, becoming deeply longitudinally furrowed when dry.

HABITAT AND ROLE: *Ceratodon purpureus* is a cosmopolitan weedy, gregarious species that is common on disturbed sites such as road edges, banks, tracks, the soil of upturned tree roots, etc. When found with sporophytes the usually purple-red setae, especially when backlit, are particularly striking. Like *Funaria hygrometrica, Ceratodon purpureus* is an early colonist of burnt ground but is replaced within a few years by a sequence of other bryophytes and flowering plants.

FAMILY: Climaciaceae

GENUS AND SPECIES: *Climacium dendroides* (Hedw.) F. Weber & D. Mohr (Fig. 209)

Figure 209. *Climacium dendroides* (Hedw.) F. Weber & D. Mohr

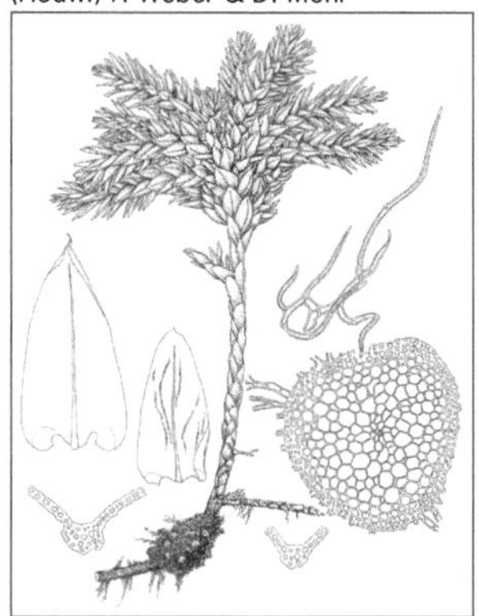

COMMON NAME: Tree climacium moss

STRIKING FIELD CHARACTERS: Plants resemble tiny trees with an apical cluster of diverging branches.

MACRO AND MICRO DESCRIPTION: Plants dull dark green to yellowish green or yellowish brown; dendroid, erect branched secondary stems arising from a horizontal, tomentose primary stem. Secondary stems erect, 2–10 cm high, frequently branching from near the top of the stem. Branch leaves erect, imbricate, plicate, 2–3 mm long, about 1 mm wide, oblong-lanceolate to lingulate (tongue-shaped), the apex obtuse and margins coarsely toothed, elsewhere the margins serrulate to the base; costa strong below, weaker above, reaching nearly to the leaf apex. Setae flexuose, to 3 cm long; capsule erect, symmetric, cylindric, 2–3 mm long, yellow-brown to brown. Rarely found fruiting.

HABITAT AND ROLE: This is a striking dendroid (refers to the tree-like growth form) species of wetter habitats. There is a good deal of variation in morphology, with shorter stems and more crowded branches being found in plants from drier habitats.

FAMILY: Dicranaceae

GENUS AND SPECIES: *Dicranum acutifolium* (Lindb. & Arn.) C. Jens. (Fig. 210)

COMMON NAME: Acuteleaf dicranum moss

Figure 210. *Dicranum acutifolium* (Lindb. & Arnell) C. Jens.

STRIKING FIELD CHARACTERS: The species is recognized by its erect-spreading leaves that are slightly curled when dry, the distal abaxial surface of the costa being smooth to weakly papillose and not ribbed, and the narrow acute leaf apices. Other key identifying and differentiating features require microscopic examination.

MACRO AND MICRO DESCRIPTION: Plants light green to light brownish green, in dense to loose tufts. Stems 2–7 cm long, branched above, tomentose with reddish brown rhizoids. Leaves lanceolate, erect-spreading, concave below, keeled above, narrowly acute to acuminate; margins serrulate above; costa percurrent to shortly excurrent, smooth to weakly papillose abaxially towards the apex. Setae solitary, 1.5–2.5 cm long, yellow to reddish yellow; capsules brown to reddish brown, 2–3 mm long, arcuate, inclined to horizontal, furrowed when dry.

HABITAT AND ROLE: This is an arctic-alpine species that occurs on soil, rock, humus, or rotting wood.

FAMILY: Dicranaceae

GENUS AND SPECIES: *Dicranum polysetum* Swartz (Fig. 211)

COMMON NAME: Dicranum moss

Figure 211. *Dicranum polysetum* Swartz

STRIKING FIELD CHARACTERS: The strongly undulate leaves coupled with the leaf margins being recurved below and strongly toothed above, the dense whitish to brownish red tomentum on the stems, the long stems, and usually aggregated sporophytes are useful identifying characteristics. Further differentiation from other *Dicranum* species requires microscopic identification.

MACRO AND MICRO DESCRIPTION: Plants in light green, loose, glossy tufts. Stems 4–10 cm or more long, densely tomentose with white to brownish rhizoids. Leaves strongly undulate, 5–10 mm long, lanceolate, concave below and keeled above, acute; margins strongly toothed in the upper half; costa ending shortly below the apex, with two toothed ridges on the abaxial surface towards the leaf apex. Setae 1.5–4 cm long, brown to reddish brown, 1–6 per perichaetium; capsules yellowish brown, 2–3.5 mm long, arcuate, inclined to horizontal, furrowed when dry.

HABITAT AND ROLE: Occurs on humus, soil over rock, on decaying wood, in deciduous and conifer forests and woodland.

FAMILY: Dicranaceae

GENUS AND SPECIES: *Dicranum scoparium* Hedw. (Fig. 212)

COMMON NAME: Dicranum moss

STRIKING FIELD CHARACTERS: This is one of the most variable *Dicranum* species in the area. Particularly confusing is the variation in leaf habit, from strongly falcate-secund (curved and pointing to one side of the stem), to straight and erect. The leaves are smooth, but may be slightly undulate, particularly towards the upper part of the leaf, leading to possible confusion with *Dicranum polysetum*. The upper leaf margin may be weakly to strongly toothed. The abaxial surface of the distal part of the costa usually has 2–4 ridges, but these are not always distinct, even with microscopic examination. Microscopic examination is usually needed for certain identification, particularly in more northerly or higher altitude habitats.

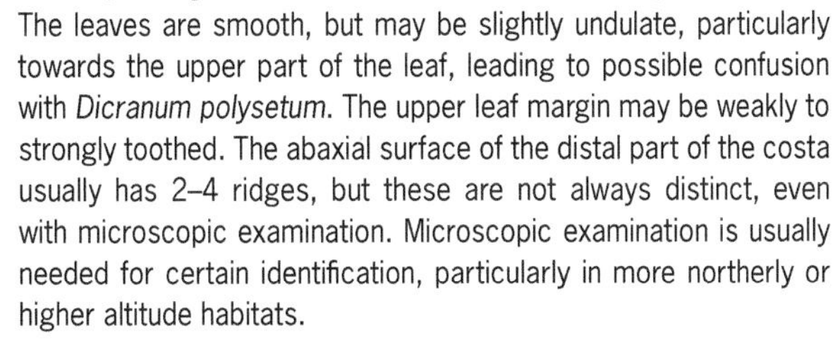

Figure 212. *Dicranum scoparium* Hedw.

MACRO AND MICRO DESCRIPTION: Plants form loose, light to dark green, glossy or dull tufts. Stems 2–10 cm tall, tomentose with white to brown rhizoids. Leaves lanceolate, very variable in habit, from falcate-secund to straight and erect, 4–10 mm long, concave below, keeled above, apex acute to obtuse; margins usually strongly serrate towards the apex; costa ceasing shortly below the apex to shortly excurrent, usually with two to four toothed ridges on abaxial surface towards the apex; lamina cells elongate with pitted cell walls. Setae solitary, 2–4 cm long, yellowish- to reddish-brown; capsules yellowish-brown, 2.5–4 mm, arcuate, inclined to horizontal, smooth to striate when dry.

HABITAT AND ROLE: Occurs on soil, humus, decaying wood, tree bases, in dry to mesic woodlands and sometimes in wetlands.

FAMILY: Ditrichaceae

GENUS AND SPECIES: *Distichium capillaceum* (Hedw.) Bruch & Schimp. (Fig. 213)

COMMON NAME: Distichium moss

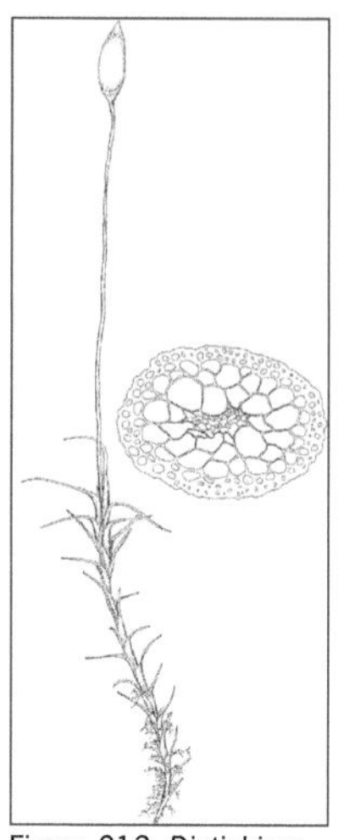

Figure 213. *Distichium capillaceum* (Hedw.) Bruch & Schimp.

STRIKING FIELD CHARACTERS: The two rows of leaves and stems with numerous rhizoids are characteristic. It appears to be more widespread than a similar species, *Distichium inclinatum* (Hedw.) Bruch, Schimp., & W. Guembel, which is more northerly and alpine in distribution. The capsule of *Distichium inclinatum* is inclined, not erect, and the leaves are more stiffly erect. Other differences are found at the microscopic level, in the spores and the structure of the peristome teeth.

MACRO AND MICRO DESCRIPTION: Plants forming loose to compact tufts, green above, reddish-brown below. Stems very variable in length, with numerous brown rhizoids matting the lower stems together. Leaves in two rows; from an elongate sheathing base contracted sharply to an elongate, spreading to squarrose, papillose to roughened subula; costa extending to the leaf apex. Setae erect, to 2 cm long; capsules straight and erect or nearly so, ovoid-cylindrical, brown.

HABITAT AND ROLE: The species is often found on calcareous soils and rocks.

FAMILY: Ditrichaceae

GENUS AND SPECIES: *Ditrichum flexicaule* (Schwaegr.) Hampe (Fig. 214)

COMMON NAME: Ditrichum moss

STRIKING FIELD CHARACTERS: The usually dense tufts, stems that are tomentose below, stiff leaves with the subula about equal in length to the sheathing base, are all useful field-identifying characters.

Figure 214. *Ditrichum flexicaule* (Schwaegr.) Hampe

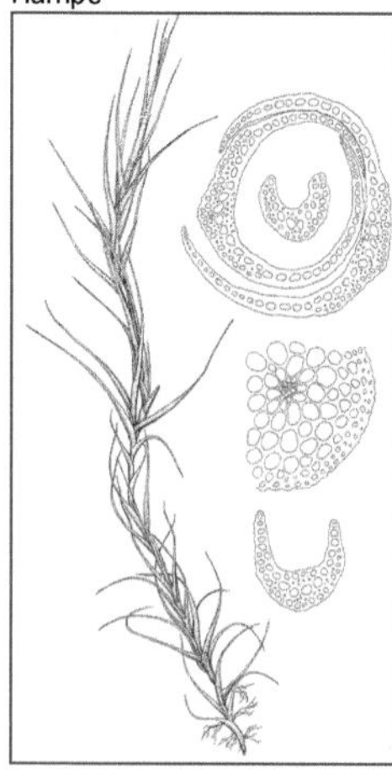

MACRO AND MICRO DESCRIPTION: Plants in dense tufts, green to brownish-green. Stems to 4 cm long, usually tomentose below, sometimes only slightly so. Leaves stiff to flexuose, rarely falcate, to about 3 mm long, from an ovate to elongate-ovate sheathing base contracted sharply to a stiff subula; margins entire or weakly denticulate near the apex; costa long-excurrent. Setae dark red-brown, flexuose, to 2 cm long; capsules erect, dark brown, cylindric, to 1.5 mm long.

HABITAT AND ROLE: This is a highly polymorphic species that occurs primarily on calcareous substrates, often in dry or exposed sites. *Ditrichum flexicaule* sensu lato has usually included the long-leaved and larger forms that have been referred to *Ditrichum crispatissimum* (now placed as a synonym of *Ditrichum gracile* (Mitt.) O.Kuntze). While often found infertile, *Ditrichum flexicaule* is more commonly found with sporophytes than the long-leaved *Ditrichum gracile*. Anatomical examination is required to unequivocally separate these from some other dense tuft-forming mosses.

FAMILY: Amblystegiaceae

GENUS AND SPECIES: *Drepanocladus* sp. (Fig. 215a–b)

COMMON NAME: Drepanocladus moss

STRIKING FIELD CHARACTERS: This species is rarely found with capsules. When growing submerged, the leaves are often straight to only slightly curved and somewhat distant. Separation of this from other related species can be difficult without microscopic examination. The high degree of morphological variation found in different forms from different habitats has led to numerous forms, varieties, and even species being described.

MACRO AND MICRO DESCRIPTION: Plants form dark green to yellowish green or brownish mats. Shoots very variable in length, to 25 cm long; stems spreading to erect when crowded together; branching irregularly pinnate. Leaves not plicate, falcate-secund (strongly curved and pointing to one side of the stem), sometimes more or less straight to slightly curved, ovate-lanceolate to lanceolate, 2–4 mm long, gradually narrowed to a point, the base somewhat cordate, slightly decurrent at the angles; margins plane, and entire; costa narrow, reaching one-half to three-fourths length of the leaf; lamina cells narrow elongate-flexuose above, shorter and wider and more rectangular below, those of the basal angles hyaline to yellow-brown, inflated. Setae 3–6 cm long; capsules 2–3 mm long, curved.

Figure 215a–b. *Drepanocladus* sp.

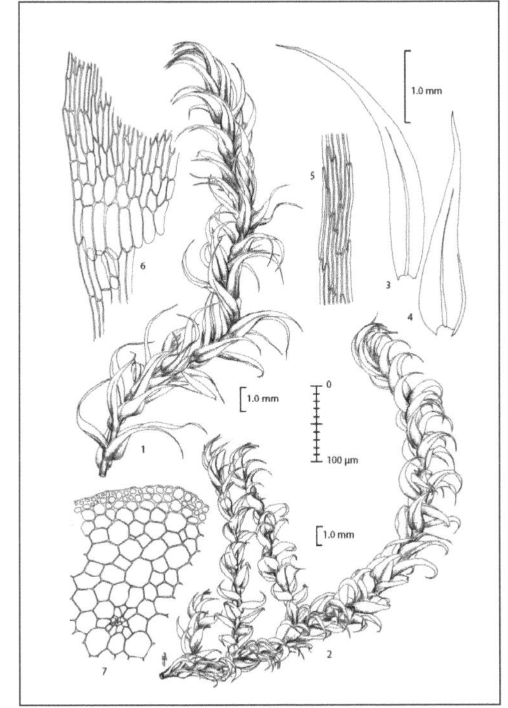

HABITAT AND ROLE: It is a moss of wet habitats, often being submerged. *Drepanocladus* is closely related to *Sanionia*, but differs in the leaves not being plicate, the stems in section lacking an outer layer of enlarged hyaline cells, and the wetter habitat.

FAMILY: Hylocomiaceae

GENUS AND SPECIES: *Hylocomium splendens* (Hedw.) Schimp. (Fig. 216a–b)

COMMON NAME: Stair-step feather moss

STRIKING FIELD CHARACTERS: *Hylocomium splendens* is commonly called the "stair-step moss" because of the characteristic manner of producing annual layers of growth. There is no other moss with the same growth pattern. Its frondlike shoots are both attractive and distinctive. It often grows in association with another feather moss, the red-stemmed and pinnate *Pleurozium schreberi* (Brid.) Mitt.

Figure 216a–b. *Hylocomium splendens* (Hedw.) Bruch, Schimp. & Guembel

MACRO AND MICRO DESCRIPTION: Plants robust, forming loose yellowish green to olive-green, dull to shiny mats. Stems stiff, spreading to ascending, curved, irregularly pinnate to bipinnate and frondlike; successive fronds produced step-wise, originating as an arched innovation from the middle and dorsal side of the previous year's growth. Stems and branches reddish, with abundant branched paraphyllia. Stem leaves 2–3 mm long, broadly oblong-ovate, abruptly narrowed to a point; costa double, reaching one-third to one-half the leaf length. Leaves of primary branches to 1.5 mm long, oblong-ovate, acute or acuminate; leaves of branchlets shorter, to about 0.8 mm, oblong-lanceolate, sharply or bluntly acute. Setae to 3 cm long, orange-brown to reddish, flexuose; capsules to about 2.5 mm long, brown, constricted below the mouth when dry.

HABITAT AND ROLE: This is a common moss of boreal woodlands and coniferous forests.

FAMILY: Bryaceae

GENUS AND SPECIES: *Leptobryum pyriforme* (Hedw.) Wilson (Fig. 217)

COMMON NAME: Leptobryum moss

STRIKING FIELD CHARACTERS: The narrow leaves and inclined to generally somewhat pendulous pear-shaped capsules are distinctive. The lower leaves are small, gradually increasing in length up the stem, with the upper and perichaetial leaves often

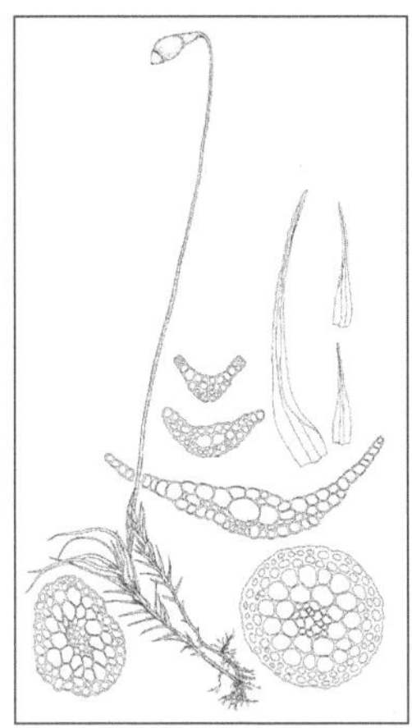
Figure 217. *Leptobryum pyriforme* (Hedw.) Wilson

forming a comal tuft at the stem tip. Pendulous pear-shaped glossy capsules are distinctive.

MACRO AND MICRO DESCRIPTION: Plants forming loose to dense patches, light green to yellowish green. Stems simple, erect, to 2 cm high. Leaves flexuose-spreading when moist; lower leaves lanceolate, acute to acuminate, upper and perichaetial leaves becoming longer and 4–5 mm in length; margins plane, unbordered, finely denticulate above; costa ceasing shortly below the apex in lower leaves, excurrent in upper leaves; lamina cells linear-rhomboidal throughout. Setae slender, flexuose, 1–4 cm long, orange-brown to reddish yellow; capsules pyriform, inclined to pendulous, brown to yellowish brown, 2–2.5 mm long.

HABITAT AND ROLE: Occurs on soil, humus, rotting wood, and in disturbed habitats.

FAMILY: Mniaceae

GENUS AND SPECIES: *Mnium thomsonii* Schimp. (Fig. 218)

Figure 218. *Mnium thomsonii* Schimp.
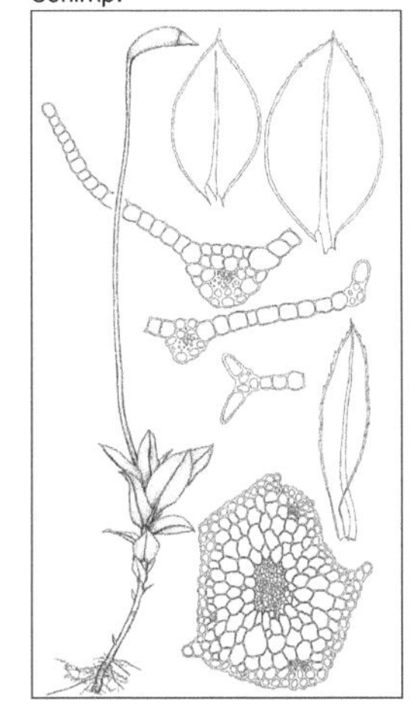

COMMON NAME: Olive-green calcareous moss

STRIKING FIELD CHARACTERS: The toothed margins of the leaves and the relatively robust size of the plants provide useful microscopic characters.

MACRO AND MICRO DESCRIPTION: Plants are found in loose green to brownish green tufts. Stems reddish brown, 2–3 cm high. Leaves contorted when dry, spreading when moist, 2–4 mm long to 1.5 mm wide, ovate to oblong-ovate, acute to acuminate, the basal margins decurrent; leaf margins with a bistratose border of narrow elongate cells, the border often reddish; margins below midleaf with double teeth; costa ceasing in the apex or just below. Setae yellowish above, reddish brown below, 2–3 cm long; capsules horizontal to pendent, yellow-brown, cylindrical, the urn to 4 mm long.

HABITAT AND ROLE: Occurs on moist soils in shaded sites, particularly on banks, and often in calcareous habitats.

FAMILY: Orthotrichaceae

GENUS AND SPECIES: *Orthotrichum speciosum* Nees in J.W. Sturm (Fig. 219)

COMMON NAME: None

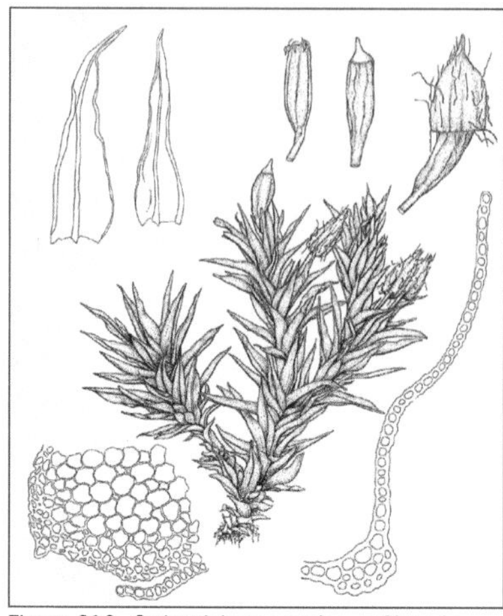

Figure 219. *Orthotrichum speciosum* Nees in J.W. Sturm

STRIKING FIELD CHARACTERS: Growing on trees and branches the loose tufts and ribbed capsules are good field identifying characters.

MACRO AND MICRO DESCRIPTION: Plants dark green to brownish green, forming loose tufts on branches and trunks of conifers and deciduous trees. Stems are 1–3 cm high and branched. Leaves loosely erect, slightly contorted, lanceolate, acuminate, 3–4 mm long, margins revolute to near the apex, entire; costa ending in the apex or just below. Setae 1–3 mm long; capsule exserted, yellow-brown, urn cylindrical to ovate-cylindric, 1.5–2.5 mm long, lightly ribbed.

HABITAT AND ROLE: This is a common species growing on conifers and deciduous trees. Some other similar *Orthotrichum* species also are found growing on trees.

FAMILY: Meesiaceae

GENUS AND SPECIES: *Paludella squarrosa* (Hedw.) Brid. (Fig. 220a–b)

Figure 220a–b. *Paludella squarrosa* (Hedw.) Brid.

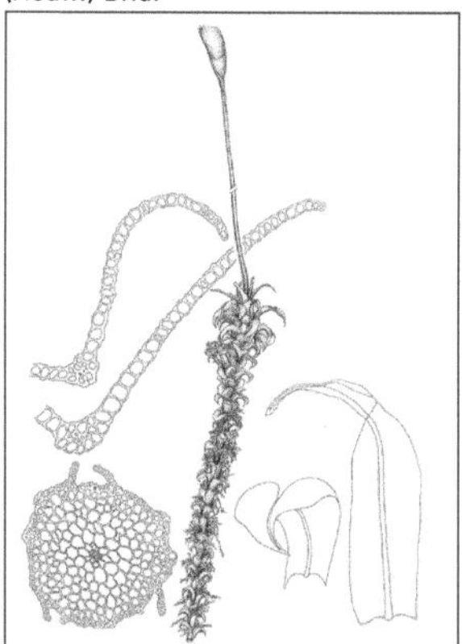

COMMON NAME: Angled paludella moss

STRIKING FIELD CHARACTERS: Not a particularly common species but the strongly squarrose-recurved leaves with papillose-mammillose (opaque) cells, and erect stems with axillary clusters of brown rhizoids are characteristic.

MACRO AND MICRO DESCRIPTION: Plants forming dark green to yellowish green tufts. Stems to 15 cm high, tomentose below, the rhizoids often clustered in the leaf axils. Leaves strongly squarrose-recurved when moist; to 2 mm long, 1 mm wide, ovate-lanceolate, broadest at the middle, acute, keeled; margins sharply serrate above, and smooth below; costa single, ceasing below the leaf apex; middle and upper leaf cells strongly mamillose-papillose. Setae to 5 cm long, sometimes longer; capsule brown, smooth, arcuate, to 3.5 mm long.

HABITAT AND ROLE: This is a plant of wet seepage or boggy areas.

FAMILY: Plagiotheciaceae

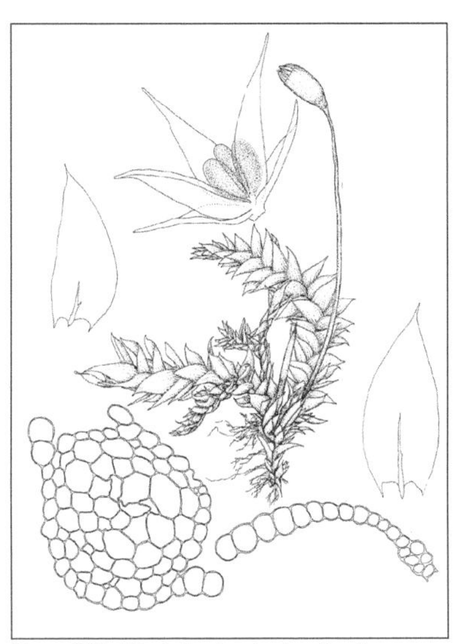

Figure 221. *Plagiothecium laetum* Bruch, Schimp. & W. Gümbel

GENUS AND SPECIES: *Plagiothecium laetum* Bruch, Schimp., & W. Gümbel (Fig. 221)

COMMON NAME: Plagiothecium moss

STRIKING FIELD CHARACTERS: The species is similar to and has been confused with *Plagiothecium denticulatum*, but that species has larger leaves, the leaf decurrencies are ovoid in outline, the cells of the decurrencies are somewhat ovoid, and the capsules are striate when dry.

MACRO AND MICRO DESCRIPTION: Plants light green to yellowish green, glossy, forming dense mats, sometimes as scattered stems. Stems to 4 cm long, prostrate, complanate-foliate (leaves flattened into one plane). Leaves imbricate to slightly distant, erect to wide-spreading, often slightly undulate, usually asymmetric, 1–2.5 mm long, 0.3–1 mm wide, oblong-ovate or ovate-lanceolate, acuminate, the margins plane, entire or occasionally weakly serrulate near the apex; leaf base decurrent, the decurrencies triangular in outline and composed of rectangular cells; costa is short and double, reaching one-fourth to one-third leaf length, sometimes absent; lamina cells narrow linear-elongate, smooth. Setae orange-brown or reddish, straight or curved, to 2 cm long; capsule light brown to orange-brown at maturity, erect to nodding, 1.5–2 mm long, smooth.

HABITAT AND ROLE: This is a woodland to wetland species found on soil, humus, rotten logs, etc.

FAMILY: Hylocomiaceae

GENUS AND SPECIES: *Pleurozium schreberi* (Brid.) Mitt. (Fig. 222a–b)

COMMON NAME: Schreber's big red stem moss; Red-stemmed feather moss

STRIKING FIELD CHARACTERS: Often forming extensive mats and often found in association with *Hylocomium splendens*, it is easily recognised by its reddish stems and branches and usually greyish tinge to the leaves. It is rarely found with capsules.

MACRO AND MICRO DESCRIPTION: Plants robust, in loose light grayish green to yellowish green mats. Stems reddish, erect-ascending, to 15 cm long, once-pinnate, the branches loosely julaceous, blunt or tapered; paraphyllia absent. Leaves loosely

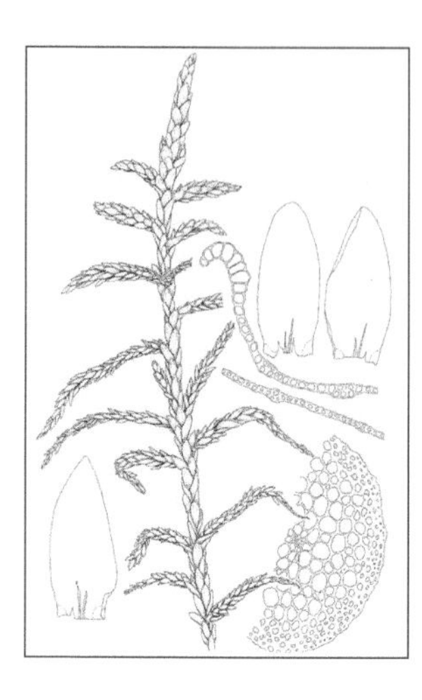

Figure 222a–b. *Pleurozium schreberi* (Brid.) Mitt.

overlapping, concave, broadly oblong-ovate, rounded at the apex; costa short and double. Branch leaves shorter and narrower than the stem leaves. Setae reddish, flexuose, to about 4 cm long; capsules about 2.5 mm long, inclined to horizontal, curved, oblong-cylindric with a tapered neck.

HABITAT AND ROLE: This is a common moss of woodlands and forests, heathland, and acidic grasslands, and occurs from drier to wetter habitats.

FAMILY: Polytrichaceae

GENUS AND SPECIES: *Polytrichum commune* Hedw. (Fig. 223)

COMMON NAME: Polytrichum moss; Hair-cap moss

STRIKING FIELD CHARACTERS: The densely hairy calyptra covering the developing capsule is responsible for the common name of "hair cap moss." This feature is also shared by other members of the Polytrichaceae, such as *Polytrichastrum.*

MACRO AND MICRO DESCRIPTION: Plants form loose tufts. Stems to 15 cm or more tall. Leaves erect and somewhat rolled when dry, spreading to recurved when moist, 6–10 mm long, from a broad, hyaline sheathing base contracted rather abruptly to a lanceolate-acuminate limb bearing numerous parallel rows of vertical lamellae along the length of the limb; margins plane to erect, sharply dentate along the limb; costa broad, excurrent as a point. Setae to 10 cm long, erect, straight to flexuose; capsules 3–5 mm long, red-brown, four-angled.

Figure 223. *Polytrichum commune* Hedw.

HABITAT AND ROLE: The species occurs in moist habitats such as spruce wetlands, shrub-heathlands, and at the margin of bogs. As for *Polytrichastrum alpinum*, the stature of the plants and leaf length varies considerably. Microscopically, the terminal or upper cell of the leaf lamellae is depressed in the middle, a distinctive characteristic. Male plants are separate, and have a series of shorter leaflike bracts forming a terminal cup, which bears the antheridia. The shoot often proliferates through the antheridial cup.

FAMILY: Polytrichaceae

GENUS AND SPECIES: *Polytrichum juniperinum* Hedw. (Fig. 224a–b)

COMMON NAME: Juniper polytrichum moss

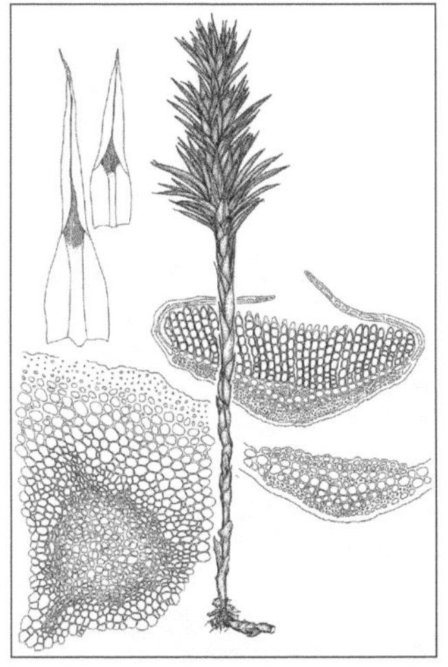

Figure 224a–b. *Polytrichum juniperinum* Hedw.

STRIKING FIELD CHARACTERS: The margins of the leaf limb are characteristically infolded over the lamellae. Microscopically, the terminal cell of the adaxial lamellae is pyriform. Male plants separate, and have a series of shorter leaflike bracts forming a terminal cup, which bears the antheridia. The shoot often proliferates through the antheridial cup.

MACRO AND MICRO DESCRIPTION: Plants form loose to dense patches. Stems to 10 cm high, often shorter. Leaves erect and appressed when dry, stiffly spreading when moist; from a sheathing base abruptly narrowed to a linear-lanceolate limb bearing numerous parallel rows of vertical lamellae along the length of the limb; margins of limb hyaline, inflexed over the adaxial surface and overlapping towards the apex of the limb; costa broad, excurrent as a short brown point. Setae to 5 cm long, reddish brown; capsules erect to suberect, rectangular, sharply four-angled, to 4 mm long.

HABITAT AND ROLE: This is a common species of well-drained soils and peats in heath, on disturbed ground, and in open woodland. *Polytrichum piliferum* Schreb. ex Hedw. is a similar species, but the leaves have long hyaline hair points, and the terminal cell of the lamellae is conic to distinctly pyriform. *P. strictum* Menzies ex Brid. is also closely similar in appearance to *P. juniperinum*, but differs in that the stems are usually covered with a whitish to light brown tomentum and the leaves are more closely appressed to the stems. The apical cell of the adaxial leaf lamellae is also pyriform in shape. The taxonomic distinction between these two species remains equivocal.

FAMILY: Hypnaceae

GENUS AND SPECIES: *Ptilium crista–castrensis* (Hedw.) De Not. (Fig. 225a–b)

Figure 225a–b. *Ptilium crista-castrensis* (Hedw.) De Not.

COMMON NAME: Knight's plume moss

STRIKING FIELD CHARACTERS: This is truly a striking moss and one of the most beautiful to be encountered. The feathery symmetrical fronds with branches curved at the tips and plicate leaves are distinctive.

MACRO AND MICRO DESCRIPTION: Plants form loose yellowish green to light green mats. Stems procumbent to erect, 5–12 cm long, regularly and closely pinnately branched, the branches in one plane and equal in length on either side of the stem, hooked at the tips. Stem leaves 2–3 mm long, plicate, ovate to ovate-lanceolate, gradually tapering to the apex, circinate-secund; margins plane, entire; branch leaves smaller than stem leaves, to 2 mm long; costa short and double or sometimes absent. Setae slender, reddish, 2.5–4.5 cm long; capsules 2–3 mm long, brown, horizontal, ovoid to oblong-ovoid, curved.

HABITAT AND ROLE: Occurring on rock, soil, humus, and rotting wood in forested areas.

FAMILY: Rhytidiaceae

GENUS AND SPECIES: *Rhytidium rugosum* (Hedw.) Kindb. (Fig. 226a–b)

COMMON NAME: Rhytidium moss

Figure 226a–b. *Rhytidium rugosum* (Hedw.) Kindb.

STRIKING FIELD CHARACTERS: The robust shoots with strongly rugose (transversely wrinkled) leaves are distinctive. Throughout its range, the species is very rarely found with capsules.

MACRO AND MICRO DESCRIPTION: Plants robust, yellow-green to golden brown, forming loose mats. Stems procumbent to ascending, irregularly pinnately branched, the branches erect to spreading. Leaves imbricate, concave, plicate and rugose, falcate-secund, 3–4 mm long, ovate-lanceolate; margins recurved, serrate towards the apex; costa single,

reaching two-third to three-fourths leaf length; lamina cells long and narrow, cells of basal angles of leaf clearly differentiated, quadrate to irregular, thick-walled. Setae 1.5–3 cm long; capsules cylindric, inclined to horizontal, arcuate, smooth, 2–2.5 mm long.

HABITAT AND ROLE: Found on soil or soil over rock, usually in calcareous habitats and often in drier places.

FAMILY: Amblystegiaceae

GENUS AND SPECIES: *Sanionia uncinata* (Hedw.) Warnst. (Fig. 227)

COMMON NAME: Sanionia moss

Figure 227. *Sanionia uncinata* (Hedw.) Warnst.

STRIKING FIELD CHARACTERS: *Sanionia* is a small genus. In appearance it is close to *Drepanocladus* but differs mainly in the leaves being strongly plicate, the stems in section having an outer epidermal layer of large hyaline and thin-walled cells, and generally occurring in drier habitat.

MACRO AND MICRO DESCRIPTION: Plants form tufts or mats, yellowish green to brownish green. Stems creeping to ascending, up to 10 cm in length, irregularly branched. Leaves plicate, usually crowded on stems and branches, 2–3 mm long, strongly falcate-circinate, narrow-lanceolate to oblong-lanceolate, gradually narrowed to a fine point from an ovoid base; margins plane, denticulate throughout, sometimes entire; costa narrow, extending to near the leaf apex; lamina cells linear-flexuose, becoming shorter near the base, those of the basal angles short-rectangular, hyaline, inflated. Setae red-brown, 2–3 cm long, flexuose; capsules 1.5–3 mm long, subcylindrical, curved, horizontal.

HABITAT AND ROLE: This is a common species of marginal wetlands, heath, and woodland. It is also commonly found as an epiphyte, especially up tree bases.

Sphagnum (Peat mosses)

There are around 89 species of *Sphagnum* included in the moss flora of North America, with 53 species known from Alaska (R. Andrus, pers. comm.), but some additional species have been described. The distribution of *Sphagnum* is worldwide but reaches its greatest abundance in the cool temperate regions of the Northern Hemisphere. It is a genus of wet, mostly acidic, habitats and is a major component of wetland vegetation. Because it is a primary component of peat deposits and has the capacity to absorb large quantities of liquid, it is important commercially.

The leafy gametophyte is unique in its structure. There are both stem leaves and branch leaves, often differing in shape and size. Mature stem leaves lack green pigmentation and are colorless. All leaves are a single layer of cells thick and lack a costa, but are uniquely differentiated into a reticulate network of alternating narrow green chloroplast-bearing cells and larger hyaline empty cells that usually have annular (ringlike) wall thickenings and whose surface has few to many pores or openings. The stems are usually differentiated into three distinct layers: an outer cortical region of relatively large thin-walled cells in one to three or four layers; an innermost region of slightly smaller thin-walled cells; and between these layers an axial cylinder (sometimes called a "wood" cylinder) of small thick-walled and often pigmented cells.

In most species, branches arise from the stem in fascicles (groups), often with divergent branches that differ in size and appearance from the pendent branches that are longer, narrower, and mostly hang down close to the stem. The capsules containing the spores are more or less round and initially green, becoming dark brown to blackish at maturity. The sporangium, unlike that of any other mosses, lacks a seta and is elevated above the perichaetium by a pseudopodium (stalk), which is an extension of the gametophyte.

Spore dispersal in *Sphagnum* is also unique amongst mosses. As the sporangium dries out it narrows, the columella collapses and is replaced by gaseous material, which exerts considerable pressure on the operculum, which is finally shed explosively with the spores.

Whereas some species of *Sphagnum* are green, many have pigmented cell walls, giving the plants a red, red-purple, orange, or brown coloration. Although impressive, color is unfortunately not an entirely reliable field guide to identification.

Sphagnum Section: Cuspidata

Plants green, whitish green to yellow-green or light brownish green. Stems green, dark brown, or occasionally pinkish; superficial cortex 0–4 layers of nonfibrillose, thin-walled cells, the outermost layer nonporose. Stem leaves triangular-lingulate, apex apiculate, acute, broad, or eroded and split. Branch fascicles with pendent branches narrower than spreading branches. Branch leaves usually longer than stem leaves, ovate to lanceolate; hyaline cells fibrillose; convex (abaxial) surface with few to numerous pores; concave (adaxial) surface aporous or porous; chlorophyllose cells triangular to trapezoid in transverse section, more broadly exposed on convex surface.

FAMILY: Sphagnaceae

GENUS AND SPECIES: *Sphagnum angustifolium* (Russow) C. Jensen (Fig. 228)

Figure 228. *Sphagnum angustifolium* (Russow) C. Jensen

COMMON NAME: Sphagnum

STRIKING FIELD CHARACTERS: Usually the branch leaves are slightly divergent to recurved, the tips standing out from the branches, giving the branches a characteristic "rough" appearance, particularly when viewed end-on. These features, together with often pinkish stems and small triangular appressed stem leaves, are useful field recognition characters.

MACRO AND MICRO DESCRIPTION: Plants relatively small, green to yellowish green to brownish, moderately stiff; capitulum strongly convex to nearly flat. Stems pale green to pale brown, sometimes pinkish, with an undifferentiated cortex. Stem leaves equilateral triangular, small, appressed to the stem, apex acute to obtuse. Branches straight to slightly curved, usually 5-ranked. Branch leaves narrowly ovate-lanceolate, straight, moderately recurved in wetter habitats.

HABITAT AND ROLE: Occurs in a wide variety of habitats in mires, fens, and muskeg.

Sphagnum section: Acutifolia

Plants small to moderate-sized; green, brown, pink, or red; with a distinct capitulum. Stems green, red, or brown; superficial cortex of two to four layers of inflated, thin-walled cells, mostly aporous. Stem leaves similar in size to branch leaves, varying from triangular to lingulate-spathulate, the apex varying from acute and entire to broad and lacerate. Branch fascicles with two to three spreading and one to two pendent branches; spreading branches usually longer and thicker. Branch leaves ovate to ovate-lanceolate; hyaline cells with round to elliptic pores along the lateral margins on convex (abaxial) surface; fewer pores on concave (adaxial) surface; chlorophyllose cells in section triangular to trapezoidal, more broadly exposed on concave (adaxial) surface.

FAMILY: Sphagnaceae

GENUS AND SPECIES: *Sphagnum fuscum* (Schimp.) H. Klinggr. (Fig. 229a–b)

COMMON NAME: Sphagnum

Figure 229a–b. *Sphagnum fuscum* (Schimp.) H. Klinggr.

STRIKING FIELD CHARACTERS: The relatively small, usually light but sometimes dark brownish plants, and the usual hummock habit are useful field recognition characters.

MACRO AND MICRO DESCRIPTION: Plants relatively small, slender, stiff, usually compact; typically deep reddish brown or greenish brown; capitulum small and flat-topped. Stems dark reddish brown. Stem leaves lingulate, apex broadly rounded and entire to lacerate. Branches long and slender to short and compact. Branch leaves ovate-lanceolate, straight, concave, apex strongly involute. Branch fascicles of two spreading and one to two pendent branches.

HABITAT AND ROLE: Often forming hummocks. A major contributor to peatland acidification throughout the Boreal region.

FAMILY: Sphagnaceae

GENUS AND SPECIES: *Sphagnum girgensohnii* Russow (Fig. 230a–b)

Figure 230a–b. *Sphagnum girgensohnii* Russow

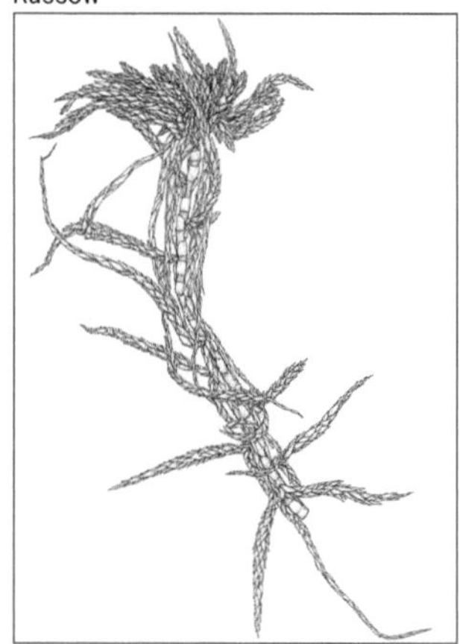

COMMON NAME: Girgensohn's sphagnum

STRIKING FIELD CHARACTERS: The characteristic wide-spreading stellate capitulum with five rays of fascicles is not always obvious. Typically, the stem leaves are only lacerate across the apex for about three-fourths of the apex width.

MACRO AND MICRO DESCRIPTION: Plants moderately sized to robust, open, stiff and slender, deep green to yellowish brown; capitulum large, flat, stellate. Stems pale green to yellow-brown. Stem leaves lingulate to lingulate-spathulate; apex broad, truncate and lacerate. Branches typically long and tapering. Branch leaves ovate to ovate-lanceolate, concave, straight, the apex strongly involute. Branches in fascicles of two spreading and one to two pendent branches.

HABITAT AND ROLE: This is a shade-tolerant carpet-forming species. *Sphagnum girgensohnii* is one of the more common *Sphagnum* species.

FAMILY: Sphagnaceae

GENUS AND SPECIES: *Sphagnum russowii* Warnst. (Fig. 231a–b)

COMMON NAME: Russow's sphagnum

STRIKING FIELD CHARACTERS: The flat, stellate capitulum, unranked branch leaves, lingulate stem leaves, and blotched red-green color are useful field characteristics.

MACRO AND MICRO DESCRIPTION: Plants of moderate size, stiff and open, or more compact in exposed sites, green or variegated red-green; capitulum flat-topped, often stellate. Stems typically mottled red and green. Stem leaves lingulate, apex broadly rounded or pointed, notched. Branches long and slender. Branch leaves ovate-lanceolate, concave, straight, apex strongly involute. Branches in fascicles of two spreading and one to two pendent branches.

HABITAT AND ROLE: This is a widespread species but one that is commonly misidentified. *Sphagnum capillifolium* (Ehrhart) Hedw. is a similar, but usually darker reddish, species, and the capitulum is rounded. The stem leaves also differ from those of *S. russowii* in that the apex is plane and rounded to acute and somewhat inrolled. *S. warnstorfii* may also be confused with *S. russowii*, but the branches usually have the leaves conspicuously 5-ranked. The red color of *S. warnstorfii* usually has a bluish cast.

Figure 231a–b. *Sphagnum russowii* Warnst.

Sphagnum Section: Squarrosa

Plants green to yellowish brown. Stems green to dark reddish brown; superficial cortex of 2–4 layers of nonfibrillose, nonporose, thin-walled cells. Stem leaves ovate, ovate-lingulate to lingulate, with a broad-fringed apex. Branch fascicles with pendent branches thinner than but about the same length as the spreading branches. Branch leaves strongly squarrose or spreading, ovate to ovate-lanceolate; hyaline cells with large, round pores at cell ends and along the lateral margins; chlorophyllose cells in transverse section ovate-triangular, elliptical to ovate-elliptic, more broadly exposed on the convex (abaxial) surface.

FAMILY: Sphagnaceae

Figure 232a–b. *Sphagnum squarrosum* Crome

GENUS AND SPECIES: *Sphagnum squarrosum* Crome (Fig. 232a–b)

COMMON NAME: Sphagnum

STRIKING FIELD CHARACTERS: Plants robust, stiff, characterized by the green to pale green to yellow-green branches having large terminal bud and strongly squarrose leaves and plants with a loose carpet-type habit.

MACRO AND MICRO DESCRIPTION: Stems green to reddish brown. Stem leaves shorter than branch leaves, ovate-lingulate to oblong-lingulate. Branches long and tapering with distinctly squarrose-spreading leaves. Branch fascicles with two spreading and two to three pendent branches.

HABITAT AND ROLE: This is a robust species. The pale green plants with the branches having strongly squarrose leaves and the loose carpet-forming habit are characteristic. It often grows in open areas, usually near ponds.

FAMILY: Splachnaceae

GENUS AND SPECIES: *Splachnum luteum* Hedw. (Fig. 233a–b)

COMMON NAME: Yellow moose and bear-dung moss

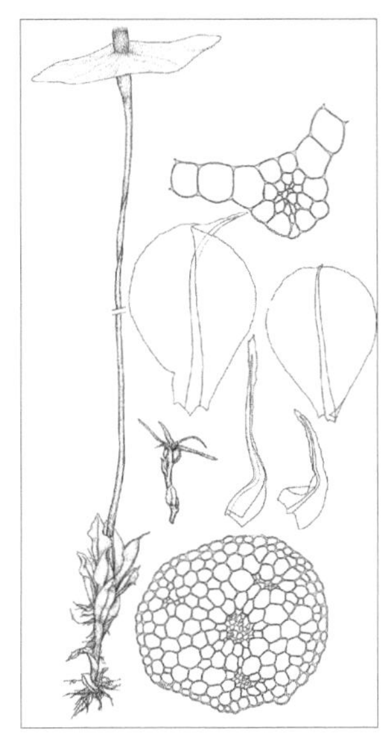

Figure 233a–b. *Splachnum luteum* Hedw.

STRIKING FIELD CHARACTERS: When fruiting, this is a truly remarkable moss. The hypophysis of the capsule becomes greatly enlarged, like a bright lemon-colored parasol, measuring up to 10 mm or more across, and it is also a bright lemon yellow color. Insects appear attracted both to the chemical attractant released as well as the color.

MACRO AND MICRO DESCRIPTION: Plants pale green to yellow-green, glossy, tomentose below, forming loose tufts on rotting dung. Stems relatively short, to 2 cm long. Leaves broadly ovate-lanceolate, acuminate, margins dentate above with short teeth, entire below; lamina cells large, thin-walled; costa ending in or slightly below the leaf apex. Seta long, to 15 cm, yellowish brown to reddish; capsules orange-brown, the hypophysis or apophysis greatly enlarged, forming a bright yellow parasol up to 10 mm diameter.

HABITAT AND ROLE: Members of the Splachnaceae are inhabitants of old dung, well-rotted wood, old bone, and other organic substrates. The spores are dispersed by insects or from a mature capsule to fresh dung. Several years elapse before the plants become reproductively mature and sporophytes are produced. The rich colors and the expanded hypophysis of both *Splachnum luteum* and *S. rubrum* provide some of the more spectacularly beautiful features to be found in the bryophytes. *S. rubrum* Hedw. is similar in appearance but the seta is a red-purple, as is the enlarged apophysis. The leaves are of similar shape to those of *S. luteum* but the margins are coarsely dentate from the apex to near the base. It appears to be a rarer species than *Splachnum luteum*. Both species occur primarily in moist woodlands.

FAMILY: Splachnaceae

GENUS AND SPECIES: *Tetraplodon angustatus* (Hedw.) Bruch, Schimp. & W. Gümbel (Fig. 234)

COMMON NAME: Toothed-leaf nitrogen moss

STRIKING FIELD CHARACTERS: When fruiting, these mosses present a striking appearance. Unlike *Splachnum*, in *Tetraplodon* the apophysis does not become greatly

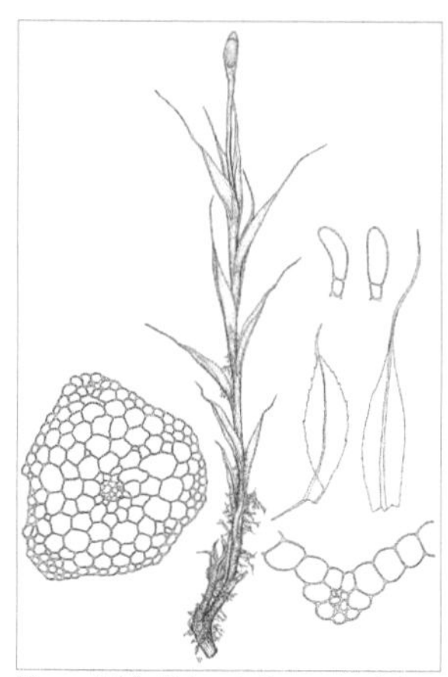

Figure 234. *Tetraplodon angustatus* (Hedw.) Bruch, Schimp. & W. Gümbel

enlarged. *T. angustatus* is similar to *T. mnioides* (see below), but the later species has considerably longer setae and larger capsules, and the leaves are shorter, not reaching to near the base of the capsule, and are in different shape.

MACRO AND MICRO DESCRIPTION: Plants form dense tufts. Stems to 4 cm long, green above, yellowish brown below. Leaves lanceolate, acuminate and with a long narrow point, 3–5 mm long, 0.5–1 mm wide; margins plane, entire or usually toothed; costa single, ending in the leaf apex. Setae 2–10 mm long; capsules to 2.5 mm long, the urn usually slightly shorter than the hypophysis.

HABITAT AND ROLE: Members of the Splachnaceae are inhabitants of old dung, well-rotted wood, old bone, and other organic substrates. **Spores** are dispersed from a mature capsule to fresh dung by insects. The leaves are large, as are the leaf cells. It usually takes several years for the plants to reach maturity, as the stems are usually densely packed and capsules abundant.

FAMILY: Splachnaceae

GENUS AND SPECIES: *Tetraplodon mnioides* (Hedw.) Bruch, Schimp. & W. Gümbel (Fig. 235)

COMMON NAME: Entire-leaf nitrogen moss

STRIKING FIELD CHARACTERS: When fruiting, these mosses present a striking appearance. Unlike *Splachnum,* in *Tetraplodon* the apophysis does not become greatly enlarged. *T. mnioides* is similar to *T. angustatus* (see above), but the this species has considerably longer setae and larger capsules, and the leaves are longer and of somewhat different shape. The reddish purple setas and nearly black mature sporangia are characteristic.

Figure 235. *Tetraplodon mnioides* (Hedw.) Bruch, Schimp. & W. Gümbel

MACRO AND MICRO DESCRIPTION: Plants form dense tufts. Stems to 8 cm long, often branched, green above, yellowish brown below. Leaves ovate-lanceolate to oblong-lanceolate, acuminate, tapering to a long narrow point; margins plane to incurved, entire; costa single, ending in the leaf apex. Setae 1.5–4 cm or more long; capsules cylindrical, 3–5 mm long, including the hypophysis, the hypophysis usually slightly wider and longer than the urn when moist.

HABITAT AND ROLE: Members of the Splachnaceae are inhabitants of old dung, well-rotted wood, old bone, and other organic substrates. **Spores** are dispersed from a mature capsule to fresh dung by insects. The leaves are large, as are the leaf cells. It usually takes several years for the plants to reach maturity, as the stems are usually densely packed and capsules abundant.

FAMILY: Brachytheciaceae

GENUS AND SPECIES: *Tomentypnum nitens* (Hedw.) Loeske (Fig. 236a–b)

COMMON NAME: Tomentypnum moss

STRIKING FIELD CHARACTERS: Plants yellowish green to golden brown, often forming extensive patches or sometimes as isolated shoots.

MACRO AND MICRO DESCRIPTION: Stems ascending to erect, 5–15 cm long, tomentose with red-brown rhizoids, irregularly branched. Leaves erect to erect-spreading, strongly plicate, triangular-lanceolate, tapering gradually to a narrow apex, with brown rhizoids arising from the proximal abaxial surface of the costa and sometimes also the lamina; stem leaves 3–4 mm long, branch leaves slightly smaller, margins entire or sinuose; costa slender, extending to three-fourths leaf or more. Setae 2–5 cm long; capsules 2–3 mm long, curved-cylindrical, horizontal. Sporophytes are infrequent.

Figure 236a–b. *Tomentyphnum nitens* (Hedw.) Loeske

HABITAT AND ROLE: This is a common and widespread species of woodland, wetlands, and fens. *Tomentypnum* is another particularly striking moss. It is an important component of the ground cover in spruce wetlands. It is a good indicator of more basic (less acid) conditions.

Description of Common Liverwort Species

FAMILY: Blepharostomaceae

GENUS AND SPECIES: *Blepharostoma trichophyllum* (L.) Dumort (Fig. 237)

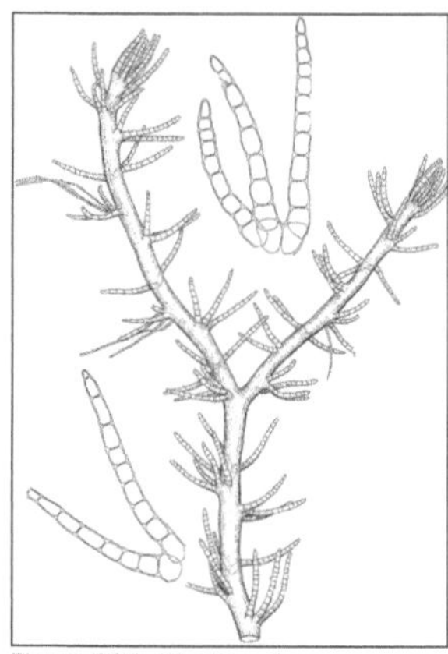

Figure 237. *Blepharostoma trichophyllum* (L.) Dumort

STRIKING FIELD CHARACTERS: Easily overlooked because of its small size. The leaves that are deeply divided into 3–4 filiform lobes are quite characteristic.

MACRO AND MICRO DESCRIPTION: Plants procumbent in loose to dense tufts or mats, or erect in short turfs amongst other bryophytes; light green to green. Stems filiform, 1–2 cm long, the leafy shoots to about 1 mm wide, simple to infrequently branched. Leaves in three similar rows, distant to imbricate, patent to suberect, often somewhat incurved and deeply divided almost to the base into three to four uniseriate, rigid lobes, the lobes with up to 14 cells.

HABITAT AND ROLE: This is a very small and delicate but characteristic species occurring in sheltered or moist habitats, on soil, litter, rotting wood, and amongst other bryophytes, often in wetlands.

FAMILY: Lophoziaceae

GENUS AND SPECIES: *Lophozia* sp. (Dum.) Dum. (Fig. 238)

STRIKING FIELD CHARACTERS: *Lophozia*, in its widest sense, is an extremely variable genus. Critical to recognizing the genus are the following features: leaves mostly bilobed, obliquely inserted to almost transverse; underleaves absent (or rarely present); gemmae mostly stellate or angular and often pigmented (brown or yellowish green); branches terminal or lateral between the leaves; free, rounded, and distally pleated perianths; and the stems somewhat dorsiventrally flattened.

Figure 238. *Lophozia* (Dum.) Dum.

MACRO AND MICRO DESCRIPTION: Plants green to dark green, reddish purple or brownish, forming carpets, tufts, or turfs; prostrate to suberect to erect. Leaves in two rows, held vertical to horizontal; transversely to obliquely inserted on the stems, usually bilobed and the lobes at least slightly asymmetrical, the postical (ventral) lobe often wider than the antical (dorsal) lobe; underleaves absent or rare. Gemmae commonly produced at the tips of leaves, stellate or irregularly angular, subquadrate to triangular or ovoid, rarely rounded.

HABITAT AND ROLE: Occurs commonly on soil in drier habitats, and on both soil and peat in woodland and wetland habitats. There are at least 28 species of *Lophozia* reported for Alaska (see footnote 1). Within the family Lophoziaceae are recognized a number of genera. Representatives of the following are found in Alaska: *Tetralophozia*

(R.M. Schuster) Schljakov, *Barbilophozia* Loeske, *Gymnocolea* (Dum.) Dum., *Eremonotus* Lindb. & Kaal. ex Pearson, *Anastrophyllum* (Spruce) Steph., and *Tritomaria* Schiffn. ex Loeske. Identification of species and genera often requires detailed microscopic examination of fresh material, particularly to note features of the oil bodies found in the leaf cells.

FAMILY: Marchantiaceae

GENUS AND SPECIES: *Marchantia latifolia* Gray (Fig. 239a–e)

STRIKING FIELD CHARACTERS: It is recognized by its nonleathery thallus, dark green color, frequently and closely forked thallus, the median longitudinal band of darker (lacking air chamber areas) and lighter (with air chambers) tissue, dentate appendages on the median ventral scales, the dorsal gemma cups with few gemmae, and the relatively long-stalked male and female receptacles.

MACRO AND MICRO DESCRIPTION: Plants form large dark green patches. Thallus 5–10 mm wide, 4–6 cm long, prostrate, branches frequent and close together. Upper surface of the thallus regularly reticulated by air-chamber areas, the medial region with darker areas lacking air chambers alternating with lighter areas containing air chambers. Ventral scales hyaline or with pale purple pigmentation, in three rows on either side of the midrib; the marginal row of scales projecting beyond the thallus margin. Male receptacle discoid, on a slender stalk 1–2 cm long, with a slightly crenulate margin. Female receptacle on a stalk 2–5 cm long; the disc developing 8–10 rays or arms alternating with the gynoecia,

Figure 239a-e. *Marchantia latifolia* Gray

the rays pendent when young, horizontal when mature; sporophytes, surrounded by a short perianth, develop between and below the rays. Discoid gemmae form in dorsal cups on the thallus, the cups having a spinulose-dentate margin.

HABITAT AND ROLE: *Marchantia latifolia* occurs in moist or humid shaded sites. A second species, *M. alpestris*, is also reported from Alaska (see footnote 1). It differs in having a rather leathery thallus, lacking the interrupted dark-light medial line, a relatively longer perichaetium, and being a species generally of higher elevations and latitudes.

FAMILY: Jungermanniaceae

GENUS AND SPECIES: *Mylia anomala* (Hook.) S. Gray (Fig. 240)

STRIKING FIELD CHARACTERS: The remarkable chartreuse-colored gemmae borne along the leaf margins and the peatland habitat, usually in *Sphagnum*, are useful field characters.

Figure 240. *Mylia anomala* (Hook.) S.Gray

MACRO AND MICRO DESCRIPTION: Plants prostrate to procumbent, ascending or erect, forming mats, or in or over bryophytes (particularly *Sphagnum*); bright green to dark green, yellowish green to brownish green or reddish brown. Shoots to 3 cm long, sometimes longer; mostly unbranched. Leaves in three rows; two rows of large suborbicular to ovoid lateral leaves and one row of small narrow-lanceolate underleaves. Lateral leaves weakly concave at the base, recurved above; margins plane to narrowly incurved distally. Gemmae one to two celled, green to yellowish green.

HABITAT AND ROLE: *Mylia anomala* is commonly found in peat mires, among or over *Sphagnum*, on moist peaty soil, in bogs and mires. It may also occur on rotting wood or on moist soils.

FAMILY: Marchantiaceae

GENUS AND SPECIES: *Preissia quadrata* (Scop.) Nees (Fig. 241)

STRIKING FIELD CHARACTERS: *Preissia* is recognized by the indistinctly reticulate (chambered) upper surface; the thallus margins developing a reddish purple pigmentation in older parts; 2 rows of purple-black overlapping ventral scales; the air chamber pores with the inner layer of epidermal cells not touching, giving rise to a cross-shaped pore opening; stalked male receptacles; the female disk with four very short lobes.

Figure 241. *Preissia quadrata* (Scop.) Nees

MACRO AND MICRO DESCRIPTION: Plants forming dull, dark grayish green prostrate patches, the older parts of the thallus sometimes reddish purple, particularly along the margins. Thallus large, to 2 cm long but often with new shoots at the apex, the lobes 5–10 mm wide; margins often somewhat undulate; air chamber areas each with acentral whitish pore surrounded by a number of superimposed rings of cells with 3–5 cells per ring, the cells of the innermost ring not touching and making the pore opening appear cruciate (cross-shaped). Male and female receptacles stalked, arising from an apical thallus notch; male receptacle disclike, female receptacle becoming hemispheric; very shallowly four-lobed, the sporangia borne on undersurface of the lobes, dehiscing by longitudinal slits.

HABITAT AND ROLE: The plants occur on moist bare ground.

FAMILY: Ptilidiaceae

GENUS AND SPECIES: *Ptilidium ciliare* (L.) Hampe (Fig. 242)

STRIKING FIELD CHARACTERS: Examination with a 10x hand lens may reveal that the anterior lamina of the leaves towards the tips of the stem and branches is clearly visible in *Ptilidium ciliare*, whereas in *P. pulcherrimum* the lamina is usually obscured or covered by overlapping cilia.

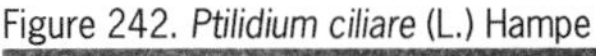

Figure 242. *Ptilidium ciliare* (L.) Hampe

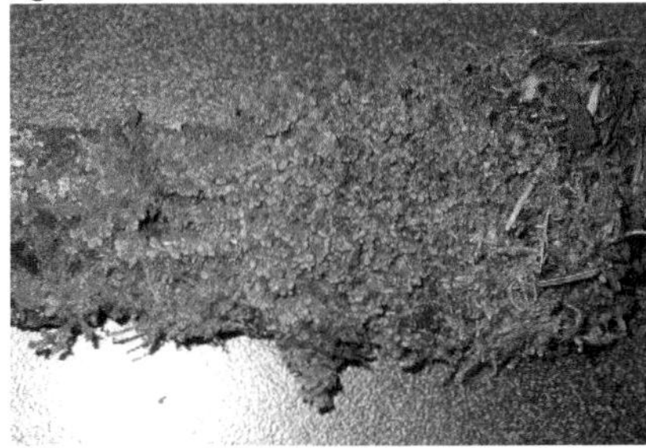

MACRO AND MICRO DESCRIPTION: Plants prostrate to suberect, in loose to dense mats, green to yellowish brown to orange-green or orange-brown, sometimes reddish; to 6 cm long, occasionally longer, the leafy shoots to 2 mm wide, occasionally slightly wider; branching pinnate to bipinnate, the secondary branches short. Leaves in three rows, two rows of larger lateral leaves and one row of smaller underleaves; leaves of main stems somewhat distant to closely imbricate, unequally bilobed and with the lobes often subdivided; leaf margins ciliate.

HABITAT AND ROLE: *Ptilidium ciliare* grows on peaty to sandy soils, on humus and litter, over rock, amongst other bryophytes, and also on bark or decaying wood. It is rarely found fertile. In the region it is easily recognized and is only likely to be confused with *P. pulcherrimum* (G.Web.) Hampe, which is a smaller species, more usually found on tree trunks and limbs and on fallen wood, and is more commonly found fertile. The leaves of *P. pulcherrimum* are more deeply divided and the marginal cilia are longer.

Glossary

abaxial—Of a leaf, the side facing away from the main shoot axis (cf. adaxial)
abruptly bulbous—An "equal" (cylindrical) stipe (stalk) giving rise to an abrupt differentiated bulbous base
acrid—Having a burning or peppery taste
acrocarpous—Of mosses, with the sporophyte produced at the apex of the main stem or branch (cf. pleurocarpous)
aculeate—Having narrow spines
acuminate—Narrowly tapered with an angle less than 45° (longer than acute)
acute—Sharp-pointed, with a terminal angle less than 90° but greater than 45°
adaxial—Of a leaf, the side adjacent to the shoot axis (cf. abaxial)
adnate—Of gills, broadly joining the stipe at right angles or "just" ascending
adnexed—Gills just barely joining the stipe, without a "notch"
agaric—A term applied to those Basidiomycetes that have gills
alar cells—Referring to cells at the basal margins or angles of a moss leaf, the cells often differing in size, shape, or color from other leaf cells
alutaceous—A light leather to pale tan color
amyloid—Staining blue with Meltzer's or Lugol's iodine
analogous—Having the same function but different structures
anamorphic fungi—Fungi producing spores (conidia) by mitosis
anastomose (-ing)—Interconnecting or running together to form an irregular network
annular zone—An often "collapsed" or evanescent partial veil or spore deposit band on the stipe
annulus—The ringlike or skirtlike structure found on the stipe of some agarics and boletes; represents a remnant of the partial veil
antheridium (pl.: antheridia)—The male gametangium of bryophytes; a multicellular cylindric to broadly globular, stalked structure containing the spermatozoids
apex—Where stipe (stalk) joins the cap
apophysis—In mosses: a strongly differentiated neck at the base of the capsule, between the top of the seta and base of the urn
apothecium (pl.: apothecia)—The cuplike or saucer-shaped fruiting body with open hymenium produced by some ascomycetes
appendiculate—Condition in which pieces of the partial veil remain attached to the margin of the cap after the latter has expanded

applanate—Flattened or expanded horizontally
appressed—With the leaves lying close to the stem
arched—Cap with down-curved margins and a depressed center (disc)
archegonium (pl.: archegonia)—The female gametangium or sex organ of bryophytes: a multicellular flask-shaped structure
arcuate—Curved like an arch
areolate—Cracked, lightly or deeply so as in drying mud
ascending—Pointing obliquely upwards, or away from the substrate
asci—Spore-producing "sacs" that also support meiosis
ascocarp—Another term for the fruiting body of an ascomycete
ascolichen—A lichen having asci and ascospores
ascomata—Fruitbodies of Ascomycetes, ascomes
ascomycetes—Fungi whose meiotic products are spores enclosed in an ascus
ascospore—The sexual spore produced within asci by fungi that are members of the ascomycetes
ascus (pl.: asci)—The saclike cell in which ascospores (usually eight) are formed; characteristic of the Ascomycetes
axil—The angle between the adaxial surface of a leaf and the stem; the angle between two adjacent branches
bacilliform—Bacillus- or rod-shaped
banded—Multiple flattened or raised "rings" on the stipe (stalk)
basidioles—Young or immature spore-producing basidia
basidiolichen—A lichen that produces basidiospores on basidia, rather than ascospores in asci
basidiome—Any fruitbody or fructification of a Basidiomycete fungus
basidiomycetes—Fungi whose meiotic products are spores produced on a basidium
basidiospore—The sexual spore produced by fungi that are members of the Basidiomycetes
basidium (pl.: basidia)—The club-shaped cell upon which basidiospores (usually four) are formed; characteristic of the Basidiomycetes
bipinnate—With both primary and secondary stems or branches pinnate
boreal—A term that refers to the high-latitude regions of the northern hemisphere dominated by coniferous forests
broad—Of gill "depth"
bulbous—Of mushroom stalks, rounded, bulblike
bullate—Having blisterlike swellings
button—A very young mushroom initial
caespitose—Aggregated or clustered; fruiting bodies occurring in groups
calcareous—Containing lime
calyptra—Of mosses, a membranous covering of the developing sporophyte
campanulate—Bell-shaped, as in the cap of some agarics
canescent—Having a "whitish downy" appearance
cap—See pileus
capillitium (pl.: capillitia)—A system of sterile threadlike elements found with the spore mass of many slime molds

capitate—With tip swollen into a head
capsule—The terminal spore-producing part of the sporophyte
cartilaginous—Being tough-pliant to tough-brittle (of stipe or stalk)
catenate—In chains
cavernous—Having hollow chambers inside the stipe (stalk)
cephalodium (pl.: cephalodia)—Small, localized raised areas on the surface of a lichen thallus that contain colonies of a blue-green alga
chalky—Brittle, breaking easily, "snapping" in half
cinereous—Bluish gray; the color of ashes
citriniform—Lemonlike in shape
clavate—Club-shaped
cleft—Of a bulbous base with a recessed sharp-walled well around the bulb top
close—Of gills, spacing and proximity to stipe (stalk), if free
clustered—Occurring close together in the same small area of the substrate
collar (-ed)—A blunt "cleft"
comal tuft—A tuft of leaves at the apex of the stem
complanate—Flattened or compressed; having leaves more or less flattened in one plane
compressed—Flattened
concave—Curved inwards, like the interior of a circle
concentric—Of rings or scaly zones arranged in circles around a common center
conchate—Shell-like or conchlike, bracketed, shelved, bivalved-shaped
concolorous—Having the same color; of one color throughout
conic—Cone-shaped, conical
conidiophore—A hyphae on which conidia are produced
conidium (pl.: conidia)—Asexual spore of anamorphic fungi
conk—Another name for a polypore; often applied to those species that occur on woody substrates
connate—Connected stems growing together, a variant of caespitose
context—The tissue that makes up the cap of a fruiting body
contorted—Irregularly curved or twisted
convex (-ed)—Curved outward or rounded outward, like the exterior of a circle
coprophilous—Dung-inhabiting; occurring on dung
cordate—Heart-shaped; a leaf with large rounded angles at the base or point of attachment
cortex—The thick covering over the spore mass in some slime molds that produce relatively large fruiting bodies; the outer part of a stem
cortina (-te)—The weblike strands that make up the partial veil of some agarics; characteristic of many members of the Cortinariaceae
costa (-te)—Of mosses, the "nerve" or "midrib" of a leaf, costate, always more than one cell thick; of liverworts, the longitudinal multistratose median region of the thallus
cottony—With long, soft, white hairs
cranioid—Having the shape of a cranium or skull
crateriform—Like a crater

crenate—Scalloped (cap margin or gill edges); with the margin edged with notches
crisped—Wavy, variously curved, twisted, or contorted
crowded—Of gills, with very little space between
crustose—A type of lichen in which the thallus is crustlike and so closely attached to the substrate as to be virtually inseparable from it, and lacking a lower cortex layer
cuboid—Being cubelike in shape
cup—A saclike, volvate stipe (stem) base
cuspidate— Eye-toothed, broadly pointed or having a sharply delineated point
cuticle—The pilipellis (skin) of the cap or of the stipe (stalk); of plants, a non-cellular coating on the outer surface of cells in contact with the environment
cyanobacterium—A blue-green alga or bacterium; in lichens, the "photobiont"
cylindric (-al)—Elongate, parallel-sided, and circular in transverse section
cystidium (pl.: cystidia)—Any number of sterile cells found in the hymenium of many Basidiomycetes; such cells usually project beyond the basidia
decurrent—With the basal margins of the leaf extending down the stem beyond the leaf insertion or attachments to the stem; of an agaric, when the attachment of the gills extends down the stipe (stalk)
decurved—Cap margins "bent" downward
deliquescent—Term applied to gills that dissolve into an "inky" liquid upon becoming mature; characteristic of many members of the genus *Coprinus*
dendroid—Like a tree; with a cluster of branches above a distinct trunklike stipe (stalk) or stem
dentate—With sharp teeth directed outwards
denticulate—Finely toothed
depressed—Of a fungal fruiting body, where the central part of the disc is "welled" or concave
detritus—Dead organic matter
dichotomous—Type of branching in which the main axis gives rise to two more or less equal secondary axes
dimidiate—Globular half circle
diploid—A cell, individual, or generation with two sets of chromosomes (2n); the typical chromosome set of the sporophyte generation
disc—The central portion of a lichen apothecium, as seen from above
distal—Away from the base or point of attachment; toward the apex of a leaf or stem
distant—Well spaced
doliiform—Barrel-formed
dorsal—Of leaves, the abaxial or lower surface; of stems or thallus, the upper surface away from the substrate
double ring—Of a "thickened" partial veil
dryads—A term applied to small alpine plants of the genus *Dryas*; these plants form ectomycorrhizae with members of several genera of Basidiomycetes
eccentric—Of the stipe (stalk) being noncentral, off-center
ectomycorrhiza—A type of mycorrhizal association in which the fungus forms a covering (called a mantle or sheath) on the outside of small roots of the host green plant

elaters—Of liverworts, elongate unicellular structures intermingled with the spores in a capsule, mostly with spirally coiled thickened bands
ellipsoid—Having the shape of an oval/ellipse, elliptic
elongate—Stretched out, linear
emarginated—Of gills, with a poorly defined notch or groove; lacking a margin
entire/even—Without teeth, more or less smooth on the margin
epigeous—Occurring above the ground
epihymenium—Upper layer of apothecium formed by apices of paraphyses over the top of the asci
epiphragm—The covering over the upper portion of the fruiting body of a bird's nest fungus; it detaches to expose the peridioles when the fruiting body is mature
equal—Straight-stemmed or gills extending from stipe (stalk) to cap margin
erect—With the leaves directed toward the stem apex; with the leaf margins curved upwards or adaxially; with the capsules straight, not curved; oriented perpendicularly to the substrate or to a horizontal axis
erectopatent—Spreading at an angle of 45° or less from the stem
eroded—Gnawed gill edges or cap margin
erose—Irregularly notched or eroded
erumpent—Bursting through the surface
eukaryotic—A type of cell that has a membrane-enclosed nucleus and other membrane-enclosed organelles
evanescent—Disappearing quickly
exciple—An outer layer of an apothecium surrounding hymenium and hypothecium
excurrent—Extending beyond the apical margin
exerted—Projecting beyond and exposed
fabaceous—With the smell of green beans/pea pods
fairy ring—An arching or circled growth pattern
falcate—Curved like a sickle
falcate-secund—Strongly curved and turned to one side
farinaceous—Smell of ground meal or cucumber
fascicle—A group, cluster, or bundle of branches (as the branches of *Sphagnum*)
fellfield—An area of tundra that is characterized by numerous rocks and sparse vegetation
ferruginous—Rusty orange to red color
fibril—Fine, fiberlike wall thickenings
fibrillose—Covered with small fibers
fibrous—Of cap or stipe (stalk) surface, composed of tough stringy tissue; of lichens, with the hyphae of the outer cortex lying parallel with the thallus axis
filamentous—Threadlike
filiform—Slender and elongate, filamentous, threadlike
fimbriate—Fringed
fistulose—Hollow, pipelike, fistular
flabelliform—Fan-shaped
flesh—Interior tissue of the fruiting body of a fungus
flexuose—Slightly and irregularly bent, or wavy

floccose—Woolly or cottony; dry and loosely arranged
flocculent—Having small tufts of fibers present
foliose—Leafy or leaflike
forked—Of gills, branching into two more or less equal arms
free—Of gills, not adjoining the stipe (stem)
friable—Breaking up, exfoliating, soon disappearing
frond—The branched or foliose part of an erect stem, including branches of a dendroid or frondose moss
fruit (ing) body—General term for the spore-producing structure produced during the reproductive stage in the life cycle of a fungus, lichen, or slime mold
fruticose—A type of lichen in which the thallus is clublike, shrublike, or more or less branched; a thallus attached to the substrate by a single point and which often develops erect branches, becoming shrublike
fugacious—Disappearing or fading early in development
fulvous—Cinnamon to reddish brown
fungoid—Earthy odor or taste
furfuraceous—Dandrufflike, flaky, scurfy
fuscous—A darkened smoky brown to purplish brown
fusiform—Spindle-shaped; tapering at both ends
gametangium—The structure bearing gametes (e.g., archegonium, antheridium)
gamete—A reproductive cell
gametophyte—Of bryophytes, the haploid (n) sexual generation–bearing antheridia and/or archegonia: the dominant generation in bryophytes
gemma (pl.: gemmae)—Asexual reproductive structures; uni- or multicellular, filamentous or variously shaped, relatively undifferentiated structures serving in vegetative reproduction
genus (pl.: genera)—A group of closely related species
glabrescent—Smooth; having the appearance of being smooth
glabrous—Smooth, bald
glandular—Dotted or punctate, spotted
glaucous—Easily rubbed-off fine white powdery "bloom"; bluish green in color or with a grayish or whitish bloom
gleba—Spore mass of a puffball and certain other related fungi
globose—Spherical; having the general form or shape of a ball
glutinous—Sticky; having the consistency of wet glue or jelly
granular—Composed of small grains, granulose
gregarious—Occurring relatively close together in the same general area of the substrate but not clustered
gynoecium—The female gametoecium, consisting of archegonia and the surrounding bracts
gyrose—Marked with wavy lines
habit—The general appearance
habitat—The local environment
hairy—Tomentose to woolly

haploid—A cell, structure, or organism having a single set of chromosomes (n); the normal chromosome level of the gametophyte generation
hemispherical—Half round
heterotrophic—Term used for an organism that is incapable of making its own organic food molecules and, as a result, must obtain them from other organisms
hirsute—Shaggy with medium stiff hairs
hispid—Stiff, straight hairs
hoary—Having a dense whitish to graying silky down
homologous—Having the same structure but different function
humic—Containing high levels of organic material derived from the partial decay of plant and animal matter, growing on humus
hyaline—Colorless or transparent
hygrophanous—Term that refers to a cap that undergoes a marked change in color as it dries out; as a result, the edge and middle of the same cap may appear very different in color
hymenium—The spore-bearing surface of a fruiting body
hymenophore—The portion of a fruiting body that bears the hymenium
hypha (pl.: hyphae)—One of the individual strands or threadlike elements that make up the fruiting body or mycelium of a fungus
hyphomycetes—Group of anamorphic fungi producing conidia not within a fruiting body
hypogeous—Occurring or growing below the ground
hypophysis—A strongly differentiated neck between the apex of the seta and the base of the urn of the capsule of mosses (cf. apophysis, neck)
hypothecium—Layer of a fruiting body below hymenium
imbricate—Closely appressed and overlapping
immarginate—Without a margin
immersed—Enclosed
inclined—Bent down; of mosses, capsules that are between the erect and the horizontal positions
incurved—Curved upward (adaxially) and inward
inferior—Toward the base; an annulus located below stipe (stalk) middle or downward hanging
inflated—Swollen, puffed up; having a somewhat rounded or swollen shape
inflexed—Bent upwards (adaxially) and weakly inwards
infundibuliform—Having a depression in the center of the cap and thus resembling a funnel
inrolled—Rolled upwards (adaxially) and tightly inwards, applied to the leaf margins
insertion—Refers to the line or point of attachment of an appendage
interascal—Between asci
intercollary—In between, middle, separating
interveinose—Veins running between the gills
involucre—Of liverworts, a short complete or incomplete tube protecting one or more antheridia or archegonia, or a developing sporophyte, replacing the perianth or rarely additional to it

involute—Rolled upwards (adaxially) and tightly inwards, applied to the leaf margins
keel—A ridge formed along a sharp fold
keeled—Sharply folded along the middle; V-shaped in cross section
labyrinthine—Mazelike pored
lacerate—Appearing as if torn or roughly cut
lacriform—Tear-shaped
lacunose—Grooved, deeply striate
lamella (pl.: lamellae)—Of a gill, one of the platelike structures upon which the hymenium occurs in agaric fungi; of mosses, parallel photosynthetic ridges or plates along a leaf blade or costa
lamellula (pl.: lamellulae)—Short gills that extend only partway to the stipe (stalk)
lamina—The flattened, generally unistratose, and green part of the leaf blade, excluding the costa and border
lamina cells—Cells of the leaf lamina
lanceolate—Lance-shaped; narrow and tapered from near the base to a long point; narrowly ovate-acuminate
lateral—At the sides
latex—Clear, wheylike, milky, or colored (yellow, orange, to violet) juice
lax—Loose; referring to large thin-walled cells as well as to the nature and spacing of leaves on a stem, or of stems in a tuft
lenticel—A small opening in bark through which gas exchange occurs
lenticular—Lens-shaped; convex on both sides and more or less circular in outline
lignicolous—High-lignin-content duff or wood decomposing
limb—The leaf blade above a differentiated leaf base
linear—Very narrow, elongate with nearly parallel sides; narrower than lingulate
lingulate—Tongue-shaped; oblong with a slightly broadened apex; long and narrow with the two longest sides more or less parallel and the distal end rounded
lobe—One of the divided distal segments or an appendage or thallus
lubricous—Moist, slippery, soapy to greasy feeling
macroscopic—Being of sufficient size to be observed with the naked eye
mamilla (pl.: mamillae)—A strongly bulging surface of a cell
mamillose—Having mamillae
marginate—Fringed
marsupium—Of liverworts, a bulbous to cylindrical structure developed at an angle to the stem and protecting the developing sporophyte, bearing rhizoids and more or less buried in the substratum
mazaedium—Dry, loose, powdery mass of free ascospores and sterile hyphae on the surface of fruiting body (characteristic of Caliciales)
mealy—Taste or smell of freshly ground flour
membranous—Thin, having the appearance of a thin layer
meristem—A localized region of growth or potential growth by cell division
mesic—Moderately moist
morphology—Form or shape
mucronate—Sharply nippled at mushroom cap center
multistratose—Of tissue, consisting of several to many layers of cells

muscicolous—Living on mosses

mycelium—System of hyphae that make up the vegetative portion of a fungus

mycobiont—The fungal component of the lichen; usually an Ascomycete (rarely a Basidiomycete)

mycology—The formal study of fungi

mycorrhiza (pl.: mycorrhizae)—A symbiotic association between a fungus and a green plant

napiform—Turniplike stipe (stalk) base; swollen or bulbous at the top and tapering abruptly toward the base

narrow—Of gills, having a thin breadth

naviculate—Boat-shaped, cymbiform

neck—Of mosses, the sterile basal portion of a capsule, sometimes considerably differentiated (cf. apophysis, hypophysis)

node—An expanded junction in the capillitium of a Myxomycete

nodulose—Spores with broad-based, blunt warts

oblique—Slanted, angled

oblong—Rectangular with rounded corners or ends

obovate—Egg-shaped, with the apex broader than the base

obovoid—An inversely ovoid solid

obpyriform—The reverse of pear-shaped

obtuse—Broadly pointed, more than 90°; sometimes referring to blunt or rounded

ochraceous—Dingy yellow to dull brownish yellow

oil body—Of liverworts, intracellular bodies containing oil globules and bound by a delicate membrane, seldom persistent on drying

opaque—Dense, impervious to light

operculum—Of mosses, a lid covering the mouth of most moss capsules

ornamentation—Any "raised" wall configuration

ostiole (-ate)—A pore that develops at the apex of a perithecium or a puffball

ovoid— Egg-shaped, with the base broader than the apex

pallid—Pale, indefinite colors

papilla (pl.: papillae)—Of cell ornamentation, a solid microscopic protuberance (cf. mamillae)

papillose (-ate)—Having papillae, papillate

parabolic—Having the form or shape of a parabola

paraphyllia—Small green outgrowths of various shapes (filiform, lanceolate, scalelike, branched), produced randomly on the stems or branches of many pleurocarpous mosses (cf. pseudoparaphyllia)

paraphysis (pl.: paraphyses)—Sterile hyphae growing upwards from the base of the hymenium among the asci

parasite—An organism that lives on or within another organism (the host), from which it obtains nutrients

parasymbiotic—Causing no visible damage to their hosts

parenchymatous—Of or having large thin-walled cells

parmelioid—Resembling lobed lichens of the genus *parmelia*

partial veil—A layer of tissue that covers the immature spore-bearing surface of some fungi
patent—Leaves spreading from the stem at an angle of 45° or more
pellucid—See-through, translucent
pellucid-striate—Gill attachments from beneath seen through a translucent cap
pendent—Hanging downwards
pendulous—Pendent or hanging
percurrent—Extending to or disappearing in the apex
perianth—Of liverworts, a tube surrounding a developing sporophyte
perichaetial leaf—A modified leaf associated with the gynoecium, collectively forming the perichaetium
perichaetium—The gynoecium; the sheathing cluster of modified leaves surrounding the archegonia
peridiole—One of the egg-shaped structures that contain basidia and basidiospores in the bird's nest fungi
peridium (pl.: peridia)—The wall of a fruiting body
perithecium (pl.: perithecia)—Sac-shaped fruiting body opening by an ostiole produced by some Ascomycetes
peristome—Of mosses, a circular structure, generally divided into 4, 8, 16, 32, or 64 teeth, arranged in a single or double row around the mouth of the capsule (absent in some mosses)
peronate—Sheathed from below, socklike
persistent—Not falling, or not deciduous; remaining for a long time
petaloid—Petal-like
photobiont—The photosynthetic component of lichens; either a green or blue-green alga
phototrophic—An organism that is capable of using light energy to make its own food molecules
phyllocladia—Of *Stereocaulon* spp., leaflike parts of a thallus growing on stem
pileus—Another name for the cap; the upper expanded portion of the fruiting body of agarics and boletes
pinnate—With numerous spreading branches on opposite side of the axis
pit—A small depression or cavity in a cell wall
pith—Central stipe (stalk) stuffing
plane—Flat, not curved or wavy, referring to the leaf margin or leaf blade
planoconvex—Term applied to a cap of a fungus that is convex but somewhat flat
plasmodium (pl.: plasmodia)—The acellular, multinucleate mass of protoplasm representing the main trophic (feeding) stage in the life cycle of a slime mold
pleurocarpous—Of mosses, producing sporophytes laterally from a perichaetial bud or short lateral branch, rather than at the stem tip (cf. acrocarpous); with the stems usually prostrate, creeping, and freely branched and growing in mats rather than tufts
plicate—With longitudinal furrows or pleats
podetium (pl.: podetia)—A hollow, upright, ascocarp-bearing structure that forms a portion of the thallus in some lichens

polymorphic—Variable, of more than one form
pore—A small aperture; in Sphagnum, a round or oval opening in the outer wall of the stem cortical cells or of a hyaline leaf cell; of liverworts, a small aperture in the dorsal epidermis of a thallus surrounded by specialized cells
porose—Having pores
procumbent—Spreading, prostrate
prokaryotic—A type of cell lacking a membrane-enclosed nucleus and other membrane-enclosed organelles
proliferations—Of *Cladonia* spp., successively developing new parts of thallus
propagule—A bud, branch, or leaf serving in vegetative reproduction
prostrate—Creeping, laying flat
protonema—Of bryophytes, a filamentous, globose, or thalloid structure arising from the germinating spore and including all stages of development up to the production of one or more gametophores (leafy shoots)
proximal—Near the base or point of attachment; the internal face of a spore (as opposed to distal)
pruina (pruinose)—A thin frosting of minute crystals covering surfaces of lichens and fruitbodies
pseudocapillitium (pl.: pseudocapillitia)—A system of irregular plates, tubes, or threadlike elements occurring within the spore mass of some slime molds
pseudoparaphyllium (pl.: pseudoparaphyllia)—Small, unistratose, filiform, or foliose structures resembling paraphyllia, but restricted to areas of a stem around a branch primordium; often found in pleurocarpous mosses
pseudopodium—A shootlike extension of the stem
pseudorhiza—Rootlike extensions
pubescent—Fine fuzzy, downy
pulverulent—Powdered; appearing as if powdered over
pulvinate—Cushion-shaped; having the general shape of a small cushion
punctuate—Dotted
pycnidium (pl.: pycnidia)—A saclike body containing conidia
pyriform—Pear-shaped
radicate (-ing)—Ramifying off into substrate, long penetrating stipe (stalk)
raphanoid—Having the odor of radish
rectangular—Shortly elongated cells with parallel sides and square-ended
recurved—Curved downwards (abaxially) and inwards; with the upper part curved away from the axis; in leaves, referring to the margins, apices, marginal teeth
reflexed—Bent backwards so that the apex points away from the axis apex; referring to the leaf margins or to leaves on a stem (as opposed to inflexed)
reticulate—Having or forming a fish net–like or ridged pattern
revolute—Rolled downwards (abaxially) and backwards, referring to a leaf margin (opposite to involute)
rhizoid—Hairlike structure that functions in water absorption and anchorage
rhizomorph—Bundled mycelial strands or cords
rhomboid—Diamond-shaped
ribbed—Of spores, longitudinally striated

rimose—Cracked, as in the cap of some agarics and boletes
rivulose—With wavy channels
rufus—Dull brick red
rugose (rugulose)—Wrinkled or corrugated
saprophyte—Organism that derives its nutrients from dead organic matter
scabers—Tufts of brown to blackish brown hyphae that occur on the stalk of some fungi, especially members of the genus *Leccinium*
scabrous—Having a rough surface, especially one with minute projections present
scrobiculate—Having glandlike shallow "pits"
scurfy—Furfuraceous, dandrufflike
seceding—Condition in which the gills pull or break away from the stipe of an agaric after being attached initially
second—Turned or curved to one side, e.g., the leaves on a stem
serrate—Saw-toothed; with the marginal teeth pointing forwards or towards the apex
serrulate—Minutely serrate
sericeous—Silky
sessile—Lacking a stalk; attached directly to the substrate
seta—The elongated portion of the sporophyte between the capsule and the foot, or point of attachment to the gametophyte plant; a bristle
sheathing—Surrounding and clasping the stem, the base of the seta, or capsule
simple—Unbranched
sinuate—Wavy, undulating, or with indentations or furrows, sinusoidal
sinuose—Wavy
sinus—A notch or indentation between two lobes
soleiform—Of spores, sole-shaped or slipper-shaped
sporangium (pl.: sporangia)—A type of fruiting body formed when a myxomycete plasmodium breaks up into a number of small portions, each of which develops into a single stalked or sessile unit; of bryophytes, the spore-bearing structure
spore—A general term for a reproductive unit in fungi, bacteria, and cryptogamic plants. Commonly one-celled, but often in fungi (and in a few bryophytes) multicelled. Usually minute, and on germination giving rise to a protonema (in bryophytes) or hyphae (in fungi).
spore print—Deposit of spores found beneath the severed cap of an agaric or bolete; used to determine the spore color in mass of a particular fungus
sporodochium (pl.: sporodochia)—Cushion-shaped mass of conidia-bearing and -supporting hyphae, characteristic of anamorphic fungi
sporophyte—The spore-bearing generation; in bryophytes, developing from the fertilized egg, remaining attached to and partially dependent on it
spreading—Forming an angle of 45° or more
squamose (squamulose)—Having squamules
squamules—Small, rounded, and somewhat leaflike lobes that form a covering over portions of the thallus in some lichens
squarrose—Spreading at right angles (90°), with the upper part bent back abruptly
squarrose-recurved—Spreading at right angles (90°), with the tips curved downwards

stalk—See stipe
stellate—Star-shaped
stipe—Stalk; the structure supporting the upper part of the fruiting body such as the cap of many agarics and boletes
stipitate—Stalked; having a stalk
striate—Marked with fine ridges or lines
strigose—Long, coarse to hairy, matted
stroma—The mass of vegetative hyphae from which the fruiting body develops in some ascomycetes
strumose—With a goiterlike swelling on one side at the base of some moss capsules
stuffed—Full, loose, cotton candy–like
sub- —A prefix meaning nearly or almost; frequently used in the sense of approaching but not quite achieving the condition in question
subdistant—Of gill spacing, separated but not widely so
subglobose—Off round
subhygrophanous—Undergoes subtle changes in color as it dries out
subperonate—A ringed swelling at the base of the stipe (stalk) in some agarics
subula—A long slender point
subulate—Slender, long-acuminate
subumbonate—Having a low raised central disc
subzonate—Concentric rings on cap not always pronounced or continuous
sulcate—Grooved
symbiosis—A situation in which two dissimilar organisms live together in an intimate association
tawny—Dull yellow-brown
terete—More or less round in cross section; cylindrical
terricolous/terrestrial—Growing on the ground
thallus (pl.: thalli)—The vegetative body of lichens or thallose and thalloid liverworts
tomentose—Woolly, with a tomentum
tomentum—A feltlike mat of fungal hyphae that covers the surface of the thallus in some lichens; the mass of hairlike filaments covering the stems of some mosses
tooth—A small, unicellular or composed of several cells, more or less triangular projection on the margin or apex
truncate—Chopped off
tuberculate—Warty or knobby
tubular (tubulose)—Tubelike, usually referring to leaves with strongly incurved or broadly overlapping leaf margins
umbilicus (umbilicate)—A centrally positioned point of attachment for the thallus that is characteristic of some rock-inhabiting lichens
umbonate—Having a central bump or raised area at the center of the cap
uncinate—Hook-notched gill attachment with a short decurrent "tooth"
underleaves—Of liverworts, the third row of leaves along the under surface of a stem or branch, often smaller or much smaller than and differing from the two rows of lateral leaves
undulate—Of a surface or margin, wavy alternately up and down

unipapillose—With single papilla per cell
universal veil—A layer of tissue that completely encloses the entire immature fruiting body of some fungi
urn—The spore-bearing portion of the capsule (as distinct from the neck)
variegate—Differently colored streaks of fibrils
veil—That which initially covers the hymenium, remaining in marginal patches or stipe fragments, the partial veil
velutinous—Velvety
venose—Having veins, veined
ventral—Of leaves, the adaxial, top, or upper surface; of stems or plants, the lower surface next to the substrate (cf. dorsal)
ventricose—Swollen or enlarged in the middle for gill depth and/or stipe (stalk)
verrucose—Having small rounded wartlike bumps
verruculose—Delicately verrucose, having low rounded bumps
villose—Long, fine hairy
vinaceous—Red wine color
viscid—Sticky to the touch; usually referring to the surface of the cap
volva—A cuplike structure found at the base of the stipe (stalk) in some fungi; represents the basal portion of the universal veil
volvate—Having a volva
warty—With small irregular surface ornamentations
waxy—Of gills belonging to *Hygrophorus*
wide-spreading—Spreading at a wide angle but less than 90°
zoned (-ate)—Concentrically ringed

Acknowledgments

A work of this nature is never the outcome of just one, but a contributory effort of many colleagues. Their contributions to each of the five major organismal sections herein presented were invaluable in bringing this volume to fruition. We acknowledge the artistic contributions made to the fungal section by **Margaret B. Hallam**, biological illustrator. Special thanks is directed to colleagues who assisted in preparing the following sections: the Macro Fungi—**Dr. Harold H. Burdsall**, USDA Forest Service, Forest Products Laboratory, Madison, WI (wood-rotting fungi); **Dr. Roland G. Treu**, University of Goroka, Goroka, Papua New Guinea (bolete and agaric fungi); and **Wayne Woodgate**, University of Alaska Fairbanks (mycorrhizae); the Lichenicolous Fungi—**Dr. Mikhail P. Zhurbenko**, Komarov Botanical Institute, St. Petersburg, Russia; the Lichens—**Dr. Linda H. Geiser**, USDA, Forest Service, Suislaw National Forest, Corvallis, OR; **Dr. Mikhail P. Zhurbenko**; **Karen L. Dillman**, USDA, Forest Service, Tongass National Forest, Petersburg, AK; and **James Riley**, USDA Forest Service, Ret., Randall, WA; the Slime Molds—**Dr. Steven L. Stephenson**, University of Arkansas; and **Dr. Yura Novozhilov**, Komarov Botanical Institute, St. Petersburg, Russia; the Mosses and Liverworts (Bryophyta)—**Olga Afonina**, Komarov Botanical Garden, St. Petersburg, Russia; **Roseann Densmore**, National Park Service, Denali Park Plant Ecologist. Thanks to **Tara Whitesell** for producing JPG images and to **Mary Elizabeth (Dierken) Laursen** for typing many of the manuscript's fungal descriptions.

This volume could not have been constructed without years of fieldwork, and we greatly appreciate the funding that has supported this long venture. Funding comes from two primary sources, the National Park Service and the USDA Forest Service.

National Park Service (NPS) funds were provided to the University of Alaska Fairbanks (UAF) through the Institute of Arctic Biology (IAB) and the USGS-BRD Alaska Fisheries Cooperative Fish and Wildlife Research Unit (ACFWRU). NPS has also provided invaluable logistical support through its Nome and Kotzebue offices with vehicle and laboratory use, fixed and rotary wing flights, and housing. We extend our appreciation to **Susan Boudreau, Steve Carwhile, Diane Chung, Lois Dalle-Molle, Greg Dudgeon, Bob Gerhardt, Gordon Olson, Peter Richter, Carl Roland, Leigh Selig**, and **Joe Van Horn**.

Support for the publication of our manuscript was provided by the U.S. Forest Service's Pacific Northwest Research Station via its Wood Utilization Research and

Development Center in Sitka, Alaska. We are particularly grateful for the assistance extended by **Allen M. Brackley, Erin Burkhart, Linda Christian, Cheryl Jennings, Bridget Kauffman, Ken A. Kilborn, Cynthia Miner, Ruth Newberry, Maria C. Stiefel, Frank Vanni, Lynn Sullivan**, and **Carolyn Wilson**.

The Australian Government Antarctic Division; the USDA Forest Services' Forest Products Laboratory, Siuslaw National Forest, and PNW Petersburg, AK; University of Arkansas, Fayetteville; the University of PNG.; and the UAF Institute of Arctic Biology have contributed logistical, clerical, secretarial, photographical, and technical support. Administrative assistance has also been faithfully provided by **Wanda Bowen** (grants), UAF–IAB; **Marta Conner** (hires), UAF–IAB; **Karen Enochs**, UAF–ACFWRU (grants); and **Heidi Harvey** (databasing), UAF–IAB. We also thank **Bess Morrison** and **Wayne Woodgate**, UAF–IAB, for field assistance.

We appreciate all editorial contributions provided by all authors, and we especially thank those who provided editorial expertise: **Dr. Robert Bandoni**, **Dr. Mariette Cole**, **Ian Herriott**, **Dr. Robert Ireland**, **Dr. Yura Novozhilov**, **Jim Riley**, **Dr. Wilf Schofield**, and **Dr. Mikhail Zhurbenko**.

English Equivalents

When you know:	*Multiply by:*	*To find:*
Celsius (°C)	1.8 and add 32	Fahrenheit
Centimeters (cm)	0.394	Inches
Hectares (ha)	2.47	Acres
Meters (m)	3.28	Feet
Micrometers (μm)	.0000394	Inches
Millimeters (mm)	.0394	Inches
Kilometers (km)	.6215	Miles

Mushroom Field and Reference Guides

Alexopoulos, C. J.; Mims, C. W. 1979. *Introductory Mycology.* New York: John Wiley & Sons. 632 p.

Ammirati, J. F.; Traquair, J. A.; Horgan, P. A. 1985. *Poisonous Mushrooms of Canada Including Other Inedible Fungi.* Markham, ON: Fitzhenry & Whiteside Limited. 396 p.

Arora, D. 1986. *Mushrooms Demystified: A Comprehensive Guide to the Fleshy Fungi.* Berkeley, CA: Ten Speed Press. 959 p.

Arora, D. 1991. *All That the Rain Promises and More: A Hip Pocket Guide to Western Mushrooms.* Berkeley, CA: Ten Speed Press. 263 p.

Atkinson, G. F. 1900. *Studies of American Fungi, Mushrooms Edible, Poisonous, Etc.* Ithaca, NY: Andrus & Church. 275 p.

Bandoni, R. J.; Szczawinski, A. F. 1964. *Guide to Common Mushrooms of British Columbia.* Victoria, BC: A. Sutton. 179 p.

Barron, G. 1999. *Mushrooms of Northeast North America: Midwest to New England.* Edmonton, AB: Lone Pine Publishing. 336 p.

Beneke, E. S. 1979. *Human mycoses.* Kalamazoo, MI: Upjohn Co. 64 p.

Bessette, A. E.; Fischer, D. W.; Bessette, A. R. 1997. *Mushrooms of Northeastern North America.* Syracuse, NY: Syracuse University Press. 784 p.

Bessette, A. E.; Miller, O. K., Jr.; Bessette, A. R.; Miller, H. H. 1995. *Mushrooms of North America in Color: A Field Guide Companion to Seldom-Illustrated Fungi.* Syracuse, NY: Syracuse University Press. 172 p.

Bigelow, H. E. 1974. *Mushroom Pocket Field Guide.* New York: Macmillan, 117 p.

Boertmann, D. 1995. *The Genus Hygrocybe.* Viborg, Denmark: Low Budget Publishing. 184 p.

Brodie, H. J. 1975. *The Bird's Nest Fungi.* Toronto, ON: University of Toronto Press. 199 p.

Charles, V. K. 1974. *Introduction to Mushroom Hunting.* New York: Dover Publishing Inc. 58 p.

Christensen, C. M. 1943. *Common Edible Mushrooms.* Minneapolis, MN: The University of Minnesota Press. 124 p.

Christensen, C. M. 1965. *Common Fleshy Fungi.* Minneapolis, MN: Burgess Publishing Co. 237 p.

Coffin, G.; Lewis, M. 1965. *Twenty Common Mushrooms and How to Cook Them.* Boston, MA: International Pocket Library. 96 p.

Cummins, G. B.; Hiratsuka, Y. 1983. *Illustrated Genera of Rust Fungi.* Rev. ed. St. Paul, MN: APS Press. 152 p.

Diederich, P., Sérusiaux, E. 2000. *The Lichens and Lichenicolous Fungi of Belgium and Luxembourg: An Annotated Checklist.* Musée National d'Histoire Naturelle, Luxembourg. 207 p.

Farr, M. L. 1981. *How to Know the True Slime Molds.* Dubuque, IA: Wm. C. Brown Co. Pub. 132 p.

Faubion, N. L. 1964. *Some Edible Mushrooms and How to Cook Them*. Portland, OR: Binfords & Mort. 198 p.

Fischer, D. W.; Bessette, A. E. 1992. *Edible Wild Mushrooms of North America: A Field-to-Kitchen Guide*. Austin, TX: University of Texas Press. 256 p.

Funk, A. 1981. *Parasitic Microfungi of Western Trees*. BC-X-222. Victoria, BC: Canadian Forestry Service, Pacific Forest Research Centre. 190 p.

Funk, A. 1985. *Foliar Fungi of Western Trees*. BC-X-265. Victoria, BC: Canadian Forestry Service, Pacific Forest Research Centre. 159 p.

Gamundi, I. J.; Horak, E. 1995. *Fungi of the Andean-Patagonian Forests: Field Guide to the Identification of the Most Common and Attractive Fungi*. Buenos Aires, Argentina: Vazquez Mazzini Editores. 141 p.

Garnweidner, E. 1996. *Mushrooms and Toadstools of Britain and Europe*. London, UK: Harper Collins Publishers. 253 p.

Graham, V. O. 1944. *Mushrooms of the Great Lakes Region*. Spec. Pub. No. 5. Chicago, IL: The Chicago Academy of Sciences. 390 p.

Groves, J. W. 1962. *Edible and Poisonous Mushrooms of Canada*. Pub. 1112. Ottawa, ON: Research Branch, Canada Department of Agriculture. 298 p.

Guba, E. F. 1970. *Wild Mushrooms Food and Poison*. Waltham, MA: Published by the author, 36 Marianne Road. 118 p.

Güssow, H. T.; Odell, W. S. 1927. *Mushrooms and Toadstools*. Ottawa, ON: F.A. Acland. 274 p.

Hanlin, R. T.; Ulloa, M. 1988. *Atlas of Introductory Mycology*. Winston-Salem, NC: Hunter Textbooks, Inc. 196 p.

Hard, M. E. 1908. *The Mushroom, Edible and Otherwise*. Columbus, OH: The New Franklin Printing Co. 609 p.

Hawksworth, D. L.; Kirk, P. M.; Sutton, B. C.; Pegler, D. N. 1995. *Ainsworth and Bisby's Dictionary of the Fungi*, 8th ed. Oxon, UK: CAB International. 650 p.

Hawksworth, D. L. 2003. The lichenicolous fungi of Great Britain and Ireland: an overview and annotated checklist. *Lichenologist 35*(3): 191–232.

Hesler, L. R. 1960. *Mushrooms of the Great Smokies*. Knoxville, TN: The University of Tennessee Press. 289 p.

Holsten, E. H.; Hennon, P. E.; Werner, R. A. 1985. *Insects and Disease of Alaskan Forests*. Reg. Rpt. 181. Juneau, AK: U.S. Department of Agriculture, Forest Service, Alaska Region. 217 p.

Hudler, G. W. 1998. *Magical Mushrooms, Mischievous Molds*. Princeton, NJ: Princeton University Press. 248 p.

Huffman, D. M.; Tiffany, L. H.; Knaphus, G. 1989. *Mushrooms and Other Fungi of the Midcontinental United States*. Ames, IA: Iowa State University Press. 326 p.

Jenkins, D. T. 1986. *Amanita of North America*. Eureka, CA: Mad River Press. 197 p.

Katsaros, P. 1989. *Illustrated Guide to Common Slime Molds*. Eureka, CA: Mad River Press. 66 p.

Krieger, L. C. C. 1936. *The Mushroom Handbook*. New York: Macmillan Publishing Co. 560 p.

Lado, C. (Ed.) 2001. *Nomenmyx: A Nomenclatural Taxabase of Myxomycetes*. Madrid, Spain: CSCI. 204 p.

Lange, M.; Hora, F. B. 1963. *A Guide to Mushrooms and Toadstools*. New York: E.P. Dutton & Co., Inc. 257 p.

Laursen, G. A.; Ammirati, J. F. 1982. *Arctic and Alpine Mycology: The First International Symposium in Arcto-Alpine Mycology*. Seattle, WA: University of Washington Press. 559 p.

Laursen, G. A.; Ammirati, J. F.; Redhead, S. A. 1987. *Arctic and Alpine Mycology II*. New York: Plenum Press. 364 p.

Laursen, G. A.; Stephenson, S. L.; Burdsall, H. H.; Densmore, R. [n.d.] *Common Mushrooms and Other Fungi of Denali National Park & Preserve*. Manuscript in preparation. On file with G. A. Laursen, University of Alaska Fairbanks, Fairbanks, AK 99775.

Lincoff, G.; Knopf, A. A. 1981. *The Audubon Society Field Guide to Mushrooms*. New York: Chanticleer Press. 926 p.

Lockwood, T. F. 2006. *Treasures from the Kingdom of Fungi*. Published by the author. 128 p.

Marshall, N. L. 1905. *The Mushroom Book*. New York: Doubleday, Page & Co. 170 p.

McCune, B.; Geiser, L. 1997. *Macrolichens of the Pacific Northwest*. Corvallis, OR: Oregon State University Press. 386 p.

McDougall, W. B. 1925. *Mushrooms*. Cambridge, MA: The Riverside Press. 151 p.

McKnight, K. H.; McKnight, V. B. 1987. *A Field Guide to Mushrooms of North America*. Boston, MA: Houghton Mifflin Company. 429 p.

McKnight, K. H.; McKnight, V. B. 1996. *Mushrooms Flash Guide*. New York: Houghton Mifflin Company. 24 p.

McIlvaine, C.; Macadam, R. K. 1902. *One Thousand American Fungi*. Rev. ed. New York: Dover. 729 p.

McKenny, M. 1971. *The Savory Wild Mushroom*. (Rev. by D.E. Stuntz.) Seattle, WA: University of Washington Press. 242 p.

Miller, O. K., Jr. 1972. *Mushrooms of North America*. New York: E.P. Dutton & Co., Inc. 360 p.

Miller, O. K., Jr.; Farr, D. 1975. An Index of the Common Fungi of North America (synonymy and common names). *Bibliotheca Mycologica 44*: 1–206.

Miller, O. K., Jr.; Miller, H. H. 2006. *North American Mushrooms: A Field Guide to Edible and Non-Edible Fungi*. Guilford, CT: FalconGuide, Globe Pequot Press. 592 p.

Orr, R. T.; Orr, D. B. 1968. *Mushrooms and Other Common Fungi of Southern California*. Berkeley, CA: University of California Press. 91 p.

Parker, H. 1994. *Alaska's Mushrooms: A Practical Guide*. Portland, OR: Alaska Northwest Books. 92 p.

Pearson, L. C. 1987. *The Mushroom Manual*. Happy Camp, CA: Naturegraph Publishers. 224 p.

Peck, C. H. 1897. *Mushrooms and Their Use*. Cambridge, MA: Cambridge Botanical Supply Co. 80 p.

Pegler, D. 1999. *The Easy Edible Mushroom Guide*. London, UK: Aurum Press Limited. 256 p.

Phillips, R. 1981. *Mushrooms and Other Fungi of Great Britain and Europe: The Most Comprehensively Illustrated Book on the Subject this Century*. London, UK: Macmillan. 288 p.

Pomerleau, R. 1951. *Mushrooms of Eastern Canada and the United States: How to Recognize and Prepare the Edible Varieties*. Montreal, QB: Chanticleer. 302 p.

Pray, L. L. 1936. *Common Mushrooms*. Botany Leaflet 18. Chicago, IL: Field Museum of Natural History. 68 p.

Progovitz, R. 2006. *Edible and Poisonous Wild Mushroom Identification Cards*. Cleveland, NY: The Forager Press. 36 cards.

Savonius, M. 1973. *All Color Book of Mushrooms and Fungi*. Hong Kong: Mandarin Publishers Limited. 72 p.

Schalkwijk-Barendsen, H. M. E. 1991. *Mushrooms of Northwest North America*. Edmonton, AB: Lone Pine Publishing. 414 p.

Seymor, J. 1978. *A Color Nature Library: Mushrooms and Toadstools*. New York: Crown Publishers Inc. 64 p.

Smith, A. H. 1949. *Mushrooms in Their Natural Habitat*. Portland, OR: Sawyer's Inc. Vol. 1. 626 p.

Smith, A. H., and Weber, N. S. 1980. *The Mushroom Hunter's Field Guide*. University of Michigan Press. 324 p.

Smith, A. H.; Smith, H. V.; Weber, N. S. 1979. *How to Know the Gilled Mushrooms*. Dubuque, IA: Wm. C. Brown Co. Pub. 334 p.

Smith, H. V.; Smith, A. H. 1973. *How to Know the Non-Gilled Fleshy Fungi*. Dubuque, IA: Wm. C. Brown Co. 402 p.

Staments, P.; Chilton, J. S. 1983. *The Mushroom Cultivator: A Practical Guide to Growing Mushrooms at Home*. Olympia, WA: Agarikon Press. 415 p.

Steinbeck, M. 1984. *Mushrooms in the Garden*. Eureka, CA: Mad River Press. 152 p.

Stephenson, S. L.; Stempen, M. 1994. *Myxomycetes: A Handbook of Slime Molds*. Portland, OR: Timber Press. 183 p.

Stevens, R.B., ed. 1981. *Mycology Guidebook*. Mycological Society of America.Seattle, WA: University of Washington Press. 703 p.

Stubbs, A. H. 1971. *Wild Mushrooms of the Central Midwest*. Lawrence, KS: The University of Kansas Press. 135 p.

Stuntz, D. E. 1977. *How to Identify Mushrooms to Genus IV: Keys to Families and Genera*. Eureka, CA: Mad River Press Inc. 94 p.

Sundberg, W. J.; Richardson, J. A. 1980. *Mushrooms and Other Fungi of Land between the Lakes*. Knoxville, TN: Tennessee Valley Authority. 39 p.

Tekiela, S.; Shanberg, K. 2006. *Start Mushrooming: The Easiest Way to Start Collecting Six Edible Wild Mushrooms*. Cambridge, MN: Adventure Publications. 128 p.

Thomas, W. S. 1948. *Field Book of Common Mushrooms*. New York: G.P. Putnam's Sons. 369 p.

Tosco, U.; Fanelli, A. 1967. *Mushrooms and Toadstools: How to Find and Identify Them*. London, UK: Orbis Publishing Limited. 102 p.

Turner, N. J.; Szczawinski, A. F. 1991. *Common Poisonous Plants and Mushrooms of North America*. Portland, OR: Timber Press. 311 p.

Tylutki, E. E. 1987. *Mushrooms of Idaho and the Pacific Northwest: Vol. 2, Non-Gilled Hymenomycetes*. Moscow, ID: University of Idaho Press. 232 p.

Vitt, D. H.; Marsh, J. E.; Bovey, R. B. 1988. *Mosses, Lichen, and Ferns of Northwest North America*. Edmonton, AB: Lone Pine Publishing. 296 p.

Watling, R. 1978. *How to Identify Mushrooms to Genus V: Cultural and Developmental Features*. Eureka, CA: Mad River Press Inc. 169 p.

Watling. R.; Watling, A. E. 1980. *A Literature Guide for Identifying Mushrooms*. Eureka, CA: Mad River Press Inc. 121 p.

Wells, M. H.; Mitchell, D. H. 1966. *Mushrooms of Colorado and Adjacent Areas*. Museum Pictorial No. 17. Denver, CO: Denver Museum of Natural History. 81 p.

References

Ahmadjian, V.; Hale, M. E. 1973. *The Lichens*. New York: Academic Press. 697 p.

Ahti, T.; DePriest, P. T. 2001. *New Combinations of Cladina Epithets in Cladonia (Ascomycotina: Cladoniaceae)*. Mycotaxon 78: 499–502.

Arora, D. 1986. *Mushrooms Demystified: A Comprehensive Guide to the Fleshy Fungi*. Berkeley, CA: Ten Speed Press. 959 p.

Boertmann, D. 1995. *The Genus Hygrocybe*. Viborg, Denmark: Low Budget Publishing. 184 p.

Brodie, H. J. 1975. *The Bird's Nest Fungi*. Toronto, ON: University of Toronto Press. 199 p.

Charles, V. K. 1974. *Introduction to Mushroom Hunting*. New York: Dover Publishing Inc. 58 p.

Coker, W. C.; Couch, J. N. 1974. *The Gastromycetes of the Eastern United States and Canada*. New York: Dover Publications, Inc. 198 p.

Courtecuisse, R.; Duhem, B. 1995. *Mushrooms and Toadstools of Britain and Europe*. London, UK: Harper Collins Publishers. 480 p.

Dennis, R. W. G. 1968. *British Ascomycetes*. Stuttgart, Germany: J. Cramer. 455 p.

Gilbertson, R. L.; Ryvarden, L. 1986. *North American Polypores: Volume 1*. Oslo, Norway: Fungiflora A/S.

Gilbertson, R. L.; Ryvarden, L. 1987. *North American Polypores: Volume 2*. Oslo, Norway: Fungiflora A/S.

Guild, B. 1977. *The Alaskan Mushroom Hunter's Guide*. Anchorage, AK: Alaska Northwest Publishing Company. 286 p.

Gulden, G.; Kolbjorn, M. J. 1988. *Arctic and Alpine Fungi 2*. Oslo, Norway: Gronland Grafiske A/S. 58 p.

Gulden, G.; Kolbjorn, M. J.; Stordal, J. 1985. *Arctic and Alpine Fungi 1*. Oslo, Norway: Gronland Grafiske A/S. 62 p.

Hale, M. E. 1979. *How to Know the Lichens*. Dubuque, IA: W.C. Brown Company Publishers. 246 p.

Holsten, E.; Hennon, P.; Trummer, L.; Shultz, M. 2001. *Insects and Diseases of Alaska Forests*. Juneau, AK: U.S. Department of Agriculture, Forest Service, Alaska Region. 242 p.

Index Fungorum. *Amanita alpina* Contu. http://www.speciesfungorum.org/Names/NamesRecord.asp?RecordID=446316.

Katsaros, P. 1989. *Illustrated Guide to Common Slime Molds*. Eureka, CA: Mad River Press. 66 p.

Lincoff, G.; Knopf, A. A. 1981. *The Audubon Society Field Guide to Mushrooms*. New York: Chanticleer Press. 926 p.

Lincoff, G.; Knopf, A. A. 1981. *The Audubon Society Field Guide to North American Mushrooms*. New York: Alfred A. Knopf, Inc. 926 p.

McCune, B.; Geiser, L. 1997. *Macrolichens of the Pacific Northwest*. Corvallis, OR: Oregon State University Press. 386 p.

McKenny, M.; Stuntz, D. E. 1987. *The New Savory Wild Mushroom*. (Rev. by J.F. Ammirati.) Seattle, WA: University of Washington Press. 250 p.

Miller, O. K., Jr. 1972. *Mushrooms of North America*. New York: E. P. Dutton & Co., Inc. 360 p.

Moser, M. 1983. *Keys to Agarics and Boleti*. Tonbridge, UK: Roger Phillips, The Whitefriars Press Ltd. 535 p.

Munsell. 1977. *Munsell Color Charts for Plant Tissues*. New Windsor, NY: Munsell Color MacBeth Division of Kollmorgen Instruments Corporation.

Ola'h, G. M. 1970. *Le genre panaeolus: essai taxinomique et physiologique*. Paris, France: Revue de Mycologie. 273 p.

Pegler, D. 1999. *The Easy Edible Mushroom Guide*. London, UK: Aurum Press Limited. 256 p.

Pojar, J.; MacKinnon, A. 1994. *Plants of the Pacific Northwest Coast: Washington, Oregon, British Columbia, and Alaska*. Vancouver, BC: Lone Pine Publishing. 527 p.

Redhead, S. A. 1977. The Genus *Mitrula* in North America. *Canadian Journal of Botany* 55: 307–325.

Schalkwijk-Barendsen, H. M. E. 1991. *Mushrooms of Northwest North America*. Edmonton, AB: Lone Pine Pubishing. 414 p.

Schumacher, T.; Kolbjorn, M. J. 1992. *Arctic and Alpine Fungi 4*. Oslo, Norway: Benjamin Sats and Trykk DA. 66 p.

Singer, R. 1975. *The Agaricales in Modern Taxonomy*. Leutershausen, Germany: J. Cramer. 912 p.

Smith, A. H. 1951. *Puffballs and Their Allies in Michigan*. Ann Arbor, MI: The University of Michigan Press. 131 p.

Smith, A. H.; Hesler, L. R. 1968. *The North American Species of Pholiota*. Monticello, NY: Lubrecht and Cramer. 402 p.

Smith, A. H.; Thiers, H. D. 1971. *The Boletes of Michigan*. Ann Arbor, MI: The University of Michigan Press. 428 p.

Thomson, J.W. 1984. *American Arctic Lichens: 1. The Macrolichens*. New York: Columbia University Press. 504 p.

Watling, R.; Gregory, N. M. 1993. *British Fungus Flora, Agarics and Boleti: 7 Cortinariaceae p.p.: Galerina, Gymnopilus, Leucocortinarius, Phyaeocollybia, Phaeogalera, Phaeolepiota, Phaeomarasmius, Pleuroflammula, Rozites, and Stagnicola*. Edinburgh, UK: Royal Botanic Garden, BPCC-AUP Aberdeen Ltd.

Ziller, W. G. 1974. *The Tree Rusts of Western Canada*. Pub. 1329. Victoria, BC: Canadian Forestry Service, Department of the Environment. 272 p.

Appendix: Mycological Reagents: Makeup and Use

Many of the important macro- and micromorphological features used in determining fungal species require chemical tests to be made. The following tables provide a list of those tests, the chemicals used in them, and some of the target fungi and their reactions to the chemistry.

Chemical	***General Use***
KOH (3 and 5%)	Tissue reviving
Melzer's reagent	Amyloidity (iodine color staining reaction)
Ethyl alcohol (70 and 95%)	Rewetting, granulations
Lactophenol cotton blue	Hyphae in plant tissues
$FeSO_4$ (10%)	Tissue
Fe_4Cl_6 (10%)	Tissue
Ferric alum (10%)	Tissue
Ferric ammonium sulfate	Granulations
Phloxine	Hyphal wall
NH_4OH	Tissue
Concentrated H_2SO_4 acid	Spore differentiation in two genera
Sulphovanillin	Cystidial
Sulfuric benaldehyde	Gloeocystidia
Gum guaiac/tannic acid	Extracellular oxidases in culture

Chemical Test	***Chemical Formulation***	***Fungal Group Use***	***Sample Target Species***	***Reaction***
KOH (3% and 5%)	3 g KOH pellets	Tissue reviving	*Xeromphalina cauticinalis*	Cherry red reaction
	97 ml distilled H_2O		*Polyporus nidulans*	Cherry red reaction
Melzer's reagent	1.5 g KI	Structural amyloidity (Iodine color staining reaction)	*Chroogomphus rutilus*	Amyloid-bluing trama
	0.5 g iodine		*Paxillus panuoides*	Pseudoamyloid (= dextrinoid): wine red spores; nonamyloid: yellowish
	20.0 g H_2O			
	22 g Chloral hydrate			
Ethyl alcohol (70% and 95%)	ETOH in H_2O	Rewetting, granulations	*Chroogomphus pileus* context	Reddish
Lactophenol cotton blue	20 g Phenol crystals	Hyphae in/on other plant tissues		
	16 ml lactic acid			
	31 ml glycerol			
	20 ml distilled H_2O			
	0.05–1 g Poirrie's (cotton) blue			
$FeSO_4$, Fe_2Cl_6, Ferric alum (10%)	10 g in 90 ml distilled H_2O	Tissue	*Ramaria*	Green tissue reaction
			Russula	Tissue reactions
			Lactarius	Tissue reactions
			Tricholoma	Tissue reactions
			Leccinum	Tissue reactions

Chemical Test	*Chemical Formulation*	*Fungal Group Use*	*Sample Target Species*	*Reaction*
Phloxine	2% aqueous soln.	Hyphal wall		Treat tissue with 70% ETOH and (sometimes in ETOH) squash in 5% KOH, then add 2% phloxine
Concn. H_2SO_4 (sulfuric acid)		Spore differentiations in two genera	*Panaeolus*	Color unchanging
			Psathyrella	Color changes: lightens/fades
Sulphovanillin	2 g pure vanilla	Cystidial	*Russula*	Cystidial bluing
	6 ml DW		*Lactarius*	Cystidial bluing
	16 ml concn. H_2SO_4			
Sulfuric benaldehyde	4.5 ml benzaldehyde	Gloeocystidia		+ reaction, blue to blue-green
	5 ml concn. H_2SO_4			
	15 ml DW			
Gum guaiac/ tannic acid	30 ml 95% ETOH	Extracellular oxidases in culture		
	0.5 g GG/TA			
	filter			
NH_4OH		Tissue		
Ferric ammonium sulfate		Granulations		

Other stains: Acid fuchsin (in lactic acid), Orange G, Picro-nigrosin, Trypan blue, Congo red, Chlorazol black, Brilliant cresol blue, Diazonium blue, Toluidine blue, lactic acid

Index to Species

Note: Botanical names are shown in italics, family names in Roman bold type, and common names in Roman type.

Italicized page numbers indicate illustrations or photographs.

A

B

C

D

E

F

G

H

I

J

K

L

M

N

O

P

Q

T

U

V

W

X

Y

About the Authors

Gary A. Laursen is with the University of Alaska Fairbanks, College of Natural Science and Mathematics, Institute of Arctic Biology, Fairbanks, AK 99775-6100; galaursen@alaska.edu.

Rodney D. Seppelt is with the Australian Antarctic Division, Channel Highway, Kingston 7050, Tasmania. This work was performed under PNW Agreement 03-GA-11261975-156; rod.seppelt@aad.gov.au.

With major contributions by

Mikhail P. Zhurbenko, Ph.D.
Laboratory for Systematics and Geography of Fungi
Komarov Botanical Institute
Russian Academy of Sciences
Prof. Popov, 2
St. Petersburg, 197376, Russia
mzhurb@gmail.com

Steven L. Stephenson, Ph.D.
Research Professor
SCEN 626
University of Arkansas
Fayetteville, AR 72701 USA
slsteph@uark.edu

Harold H. Burdsall, Jr., Ph.D.
Fungal and Decay Diagnostics, LLC
9350 Union Valley Road
Black Earth, WI 53515 USA
burdsall@fungaldecay.com

Dr. Linda H. Geiser, Ph.D.
US Forest Service
Pacific Northwest Region Air Program
P.O. Box 1148
Corvallis, OR 97339 USA
lgeiser@fs.fed.us

Roseann Van Essen Densmore, Ph.D.
US Geological Survey Alaska Science Center
4210 University Drive
Anchorage, AK 99508 USA
rdensmore@usgs.gov